AF531588

MOLECULAR MARKER AND PLANTS BIOTECHNOLOGY

MOLECULAR MARKER AND PLANTS BIOTECHNOLOGY

Dr. Manendra Kumar

RANDOM PUBLICATIONS

NEW DELHI - 110 002 (INDIA)

Molecular Marker and Plants Biotechnology

ISBN 978-93-51116-96-7

Published in 2015 in India by

RANDOM PUBLICATIONS

4376-A/4B, Gali Murari Lal, Ansari Road
New Delhi-110 002
Phone: +9111-43580356, 23289044
E-mail: randomexports@gmail.com; sales@randompublications.com; info@randompublications.com

Reprinted 2024

Type Setting by: Friends Media, Delhi-110089
Digitally Printed at : Replika Press Pvt. Ltd.

Preface

In genetics, a molecular marker (identified as genetic marker) is a fragment of DNA that is associated with a certain location within the genome. Molecular markers are used in molecular biology and biotechnology to identify a particular sequence of DNA in a pool of unknown DNA. A genetic marker is a gene or DNA sequence with a known location on a chromosome that can be used to identify individuals or species. It can be described as a variation (which may arise due to mutation or alteration in the genomic loci) that can be observed. A genetic marker may be a short DNA sequence, such as a sequence surrounding a single base-pair change (single nucleotide polymorphism, SNP), or a long one, like minisatellites. Considerable developments in biotechnology have led breeders of plants and animals to develop more efficient selection systems to replace traditional phenotype-based selection systems.

Marker assisted selection (MAS) is an indirect selection process where a trait of interest is selected, not based on the trait itself, but on a marker linked to it. For example, if MAS is being used to select individuals with disease resistance, the level of disease resistance is not quantified but rather a marker allele that is linked with disease resistance is used. The assumption is that the marker used for selection associates at high frequency with the gene or quantitative trait locus (QTL) of interest, due to genetic linkage (close proximity, on the chromosome, of the marker locus and the disease resistance-determining locus). MAS can be very useful to efficiently select for traits that are difficult or expensive to measure, exhibit low heritability, and are expressed late in development. However, it is usually essential to confirm at certain points in the breeding process that the selected individuals or their progeny do in fact express the desired phenotype or trait. The gene of interest directly causes production of protein(s) or RNA that produce a desired trait or phenotype, whereas markers (a DNA sequence or the morphological or biochemical markers produced due to that DNA) are genetically linked to the gene of interest. The gene of interest and the marker tend to move together during segregation of gametes due to their proximity on the same chromosome and concomitant reduction in recombination (chromosome crossover events) between the marker and gene of interest. For some traits, the gene of interest has been discovered and the presence of desirable alleles can be directly assayed with a high level of confidence. However, if the gene of interest is not known, markers linked to the gene of interest can still be used to select for individuals with desirable alleles of the gene of interest. When markers are used there may be some inaccurate results due to inaccurate tests for the marker. There also can be false positive results when markers are used, due to recombination between the marker of interest and gene (or QTL). A perfect marker would elicit no false positive results. The term 'perfect marker' is sometimes used when tests are performed to detect a SNP or other DNA polymorphism in the gene of interest, if that SNP or other polymorphism is the direct cause of the trait of interest. The term 'marker' is still appropriate to use when directly assaying the gene of interest, because the test of genotype is an indirect test of the trait or phenotype of interest. The process of developing new crop varieties can take almost 25 years. Now, however, biotechnology has considerably shortened the time to 7-10 years for new crop varieties to be brought to the market. One of the tools which can make it easier and faster for scientists to select plant traits is marker-assisted selection (MAS). The differences that distinguish one plant from another are encoded in the plant's genetic material, the DNA. DNA is packaged in chromosome pairs (strands of genetic material), one coming from each parent. The genes, which control a plant's characteristics, are located on specific segments of each chromosome. Together, all of a plant's genes make up its genome. Some traits, like flower colour, may be controlled by only one gene. Other more complex characteristics, however, like crop yield or starch content, may be influenced by many genes. Traditionally, plant breeders have selected plants based on their visible or measurable traits, called the phenotype. This process can be difficult, slow, influenced by the environment, and costly – not only in the development itself, but also for the economy, as farmers suffer crop losses.

This book serves not only as a textbook for students, teachers, researchers and Industries alike in plant tissue culture and biotechnology, but for general reference.

I thank all members of my team who have helped in the preparation of the book. My special thanks go to "Random Publications" who have published the book.

— *Dr. Manendra Kumar*

Contents

Chapter 1

Introduction

Biotechnology has been employed by humans for millennia; traditional applications include production of beer, cheese, and bread. But the recent developments in molecular biology have given biotechnology new meaning, new prominence, and new potential. It is this "new biotechnology" that has captured the attention of scientists, financiers, policymakers, journalists, and the public, even though the revenue it generates is as yet only a fraction of that produced by traditional biotechnology. Through the use of advanced tools, such as genetic engineering, biotechnology is expected to have a dramatic effect on the world economy over the next decade.

A variety of U.S. industries, primarily the agriculture and pharmaceutical sectors, already are profiting from this trend; an estimated $7 billion in sales were generated in 1993 as a result of the new biotechnology. These sales are expected to reach approximately $50 billion in the next decade. The United States leads the world in technological innovation and new company formation in biotechnology. The biotechnology industry is responsible for approximately 100,000 high-skill jobs generated by 1,300 biotechnology firms.

Examples of new products derived from modern biotechnology include tomatoes with an extended shelf life and a drug, tissue plasminogen activator, used to dissolve blood clots during heart attacks. But these innovations offer only a hint of the enormous potential of biotechnology. Many other new products are also in development, including disease-resistant plants, "natural" pesticides, environmental remediation technologies, biodegradable plastics, novel therapeutic agents, chemicals, and enzymes that will reduce the cost and improve the efficiency of industrial processes. A coordinated Federal research effort can provide the leverage needed to fulfil the broad promise of biotechnology, which may well play as pivotal a role in social and industrial advancement over the next 10 to 20 years as did physics and chemistry in the post-World War II period.

The Federal Role in Biotechnology

The Federal Government has propelled the development of modern biotechnology by supporting research that led to the cloning of the first gene in 1973 and the subsequent development of DNA sequencing technologies. A strong foundation has been laid that places the United States at the forefront of biotechnology research. Critical to continued innovation in all spheres of biotechnology is an ongoing Federal commitment to basic research into fundamental life processes, as well as support for the requisite infrastructure.

To date, the Federal investment in biotechnology has been focused primarily in the health field. That approximately 80 percent of the Federal biotechnology investment supports health-related projects and basic research that is broadly applicable but often health oriented. Much of this support comes from the National Institutes of Health (part of the Department of Health and Human Services), as indicated.

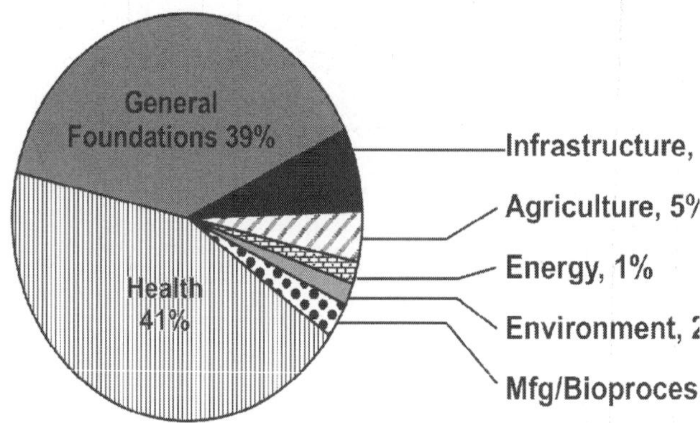

Figure: *Federal Investment in Biotechnology Research (by Area).*

The results of health-related biotechnology research are having a profound impact on medicine and health care, providing improved approaches to the diagnosis, treatment, and prevention of disease. While health-related research must remain a national priority, researchers are poised to build on the foundation in basic science to bring the power of biotechnology to bear in other fields. Only 12 percent of the Federal investment in biotechnology research supports applications outside the health arena. This report presents the overall Federal perspective on biotechnology issues, opportunities, and priorities in areas beyond health that need to be addressed by funding agencies and the research community. The report is the result of a planning effort by the Biotechnology Research Subcommittee (BRS) of the Committee on Fundamental Science of the National Science and Technology Council (NSTC). The BRS is an interagency committee charged with coordinating the Federal investment in biotechnology.

Federal Investment in Biotechnology Research

Millions of dollars *

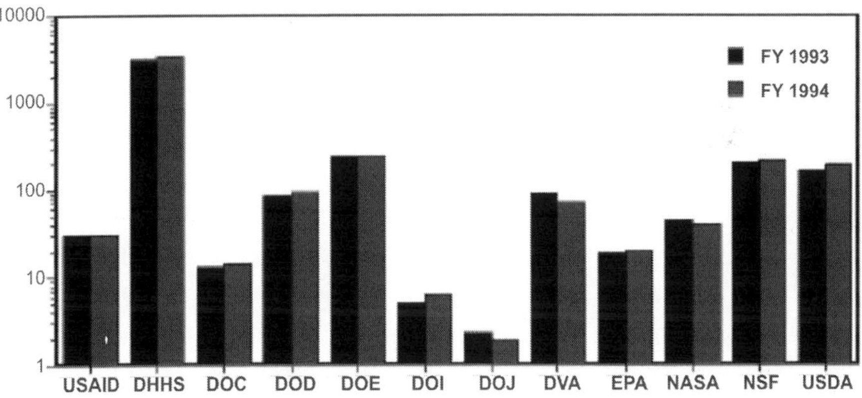

* Agency Investments are displayed using a semilogarithmic scale.

FY 1993: 4,268.7 M
FY 1994: 4,299.3 M

Figure: *Federal Investment in Biotechnology Research.*

Because biotechnology research is broadly applicable to diverse government missions and goals, it is supported by 13 Federal departments and agencies, all represented on the BRS:

Agency for International Development (USAID)

Department of Agriculture (USDA)

Department of Commerce (DOC)

Department of Defence (DOD)

Department of Energy (DOE)

Department of Health and Human Services (DHHS)

Department of the Interior (DOI)

Department of Justice (DOJ)

Department of State (DOS)

Department of Veteran Affairs (DVA)

Environmental Protection Agency (EPA)

National Aeronautics and Space Administration (NASA)

National Science Foundation (NSF)

This report is aimed at a wide audience, including the U.S. Congress, Executive Branch officials, industry, and academic institutions.

The Federal Government is one of three partners, along with the industrial and academic communities, in the collaborative venture that is biotechnology research and development. The government's role is to sponsor research that will generate new knowledge and to ensure that technological advancement contributes to the overall public good and national welfare.

Support of basic research generates innovations, from which previously unanticipated products may arise. The private sector completes the technology development cycle and ensures distribution of beneficial products to consumers.

Report Structure

The resources of Federal agencies, as well as expertise from the private sector, were brought together to identify opportunities in four rapidly developing areas of biotechnology research. Each area is addressed in a separate chapter of this report, which identifies priorities for Federal investment and summarizes current directions and specific research opportunities. (Related activities of each BRS agency are summarized in the Appendix.)

The chapters emphasize multidisciplinary, long-term research, which will be essential in order to enhance the efficacy of existing biotechnologies and develop new ones. The four areas highlighted are agricultural biotechnology, environmental biotechnology, manufacturing/bioprocessing, and marine biotechnology and aquaculture.

The concluding chapter of the report outlines infrastructure needs — facilities, human resources, databases, and mechanisms for collaboration and information exchange. A sophisticated infrastructure is required to support all areas of biotechnology research.

Based on the common themes identified in the four areas, the BRS identified three overarching priorities for Federal investment in biotechnology research. Pursuit of these priorities will ensure the development of a strong research base, which in turn will facilitate product commercialization and help maintain U.S. leadership in biotechnology.

Opportunities in Agriculture

Biotechnology offers efficient and cost-effective means to produce an array of novel, value-added products and tools. It has the potential to increase food production, reduce the dependency of agriculture on chemicals, lower the cost of raw materials, and reduce the negative environmental impacts associated with traditional production methods. In addition, the new knowledge gained through basic research into the nature of life and ecosystems at the molecular level can lead to improved farming practices and diagnostic tools for use in agriculture.

Agricultural biotechnology has the potential to produce billions of dollars in revenue per year in the next century. It will play a crucial role in promoting the nation's economic growth, improving environmental quality, and assuring innovative scientific research. Federal support for biotechnology research in this area is essential, in order to build the broad knowledge base needed to commercialize new and improved agricultural products and tools.

The Promise of Agricultural Biotechnology

Due in large part to scientific advances in crop breeding and farming techniques, world food production has doubled since 1960, and productivity from agricultural land and water usage has tripled. Today, the peoples of the world farm an area about the size of South America; without the scientific advances of the past 30 years, farmland equivalent to the entire Western hemisphere would be required.

But a dilemma lies ahead. The world's population is expected to double by the year 2030 to 12 billion, and it is not clear whether current food production is keeping pace with population growth. The question is how best to feed billions of additional people without destroying much of the planet in the process. The disappearance of tropical rain forests, wetlands, and other vital habitats will accelerate unless agriculture somehow becomes more productive and less taxing to the environment.

It seems certain that agricultural biotechnology will play a major role in resolving this dilemma. Biotechnology can be employed to improve the quality of seed grains; increase protein levels in forage crops; and instill in crops resistance to disease, insects, and viruses, as well as tolerance for droughts, floods, and extreme temperatures. In addition, biotechnology can make foods healthier and more nutritious.

For example, tomatoes and other fruits and vegetables containing increased levels of certain nutrients, such as vitamins C and E and beta carotene, may help protect against risk of chronic diseases, such as some cancers and heart disease.

Although major genetic improvements have been made in crops, progress in conventional breeding programs has been slow. Moreover, most crops grown in the United States produce less than 50 percent of their genetic potential. These shortfalls in yield are due in large part to the inability of crops to tolerate or adapt to environmental stresses, pests, and disease.

For example, some of the world's highest yields of potatoes are in Idaho under irrigation, but in 1993 both quality and yield were reduced severely because of cold, wet weather and widespread frost damage during June. And some of the world's best bread wheats and malting barleys are produced in the north-central states, but in 1993 the disease Fusarium (headblight of wheat and barley) caused an estimated $1 billion in damage.

Major advances also have been made through conventional breeding and selection of livestock. Feed efficiency (1) for poultry and swine has been increased by nearly 50 percent. Milk production per cow has more than doubled since 1955. Diseases such as hog cholera and pests such as screwworm have been eradicated. Yet major diseases of livestock still go uncontrolled and some still go undiagnosed or even unrecognized.

These and many other agricultural production hazards have defied traditional solutions such as classical breeding approaches. New research is needed to address these problems, in order to minimize risks; improve the financial stability of farms, rural communities, and agribusinesses; and assure the long-term competitiveness of U.S. agriculture in the world market. These advances must be accompanied by reduced dependency on pesticides and improved environmental sustainability.

Biotechnology can compress the time frame required to translate fundamental discoveries into applications. With improved technology and knowledge about agricultural organisms, processes, and ecosystems, opportunities will emerge to produce new and improved agricultural products in an environmentally sound manner. This chapter highlights five broad priorities in agricultural biotechnology research that merit attention by Federal agencies:

- Continue mapping and sequencing of animal/plant/microbial genomes to elucidate gene function and regulation and to facilitate the discovery of new genes as a prelude to gene modification.

- Determine biochemical and genetic control mechanisms of metabolic pathways in animals, plants, and microbes that may lead to products with novel food, pharmaceutical, and industrial uses.
- Extend understanding of the biochemical and molecular basis of growth and development including structural biology of plants and animals.
- Elucidate the molecular basis of interactions of plants and animals with their physical and biological environments, as a basis for improving the organisms' health and wellbeing.
- Enhance food safety assurance methodologies, such as rapid tests for identifying chemical and biological contaminants in food and water.

This chapter identifies, within each priority area, selected areas where additional research is needed to assure a steady supply of high-quality agricultural products at reasonable prices for decades to come. The report also highlights recent scientific advances involving Federally supported research.

Gene Sequencing and Mapping

PRIORITY: Continue mapping and sequencing of animal/plant/microbial genomes to elucidate gene function and regulation, and to facilitate the discovery of new genes as a prelude to gene modification.

To locate desired genetic traits of organisms and effectively use molecular probes to identify DNA sequences, researchers need precise and relatively complete genetic maps. The combination of detailed maps and DNA sequences of genomes can provide a guide to plant and animal breeders for the development of new, efficient, and informed breeding strategies.

The Federally funded Human Genome Project is rapidly advancing gene mapping and sequencing technologies as well as data management techniques for analysing genome data from humans and other organisms. These advances have the potential to make important contributions to the genetic improvement of crops and animals. The agricultural research community is in an ideal position to apply the technical and scientific advances from the Human Genome Project to address the biology of organisms important to agriculture, aquaculture, and forestry.

Plant Genome Research

The U.S. agricultural research community has mounted a considerable effort to understand the structure, function, and regulation of genes in a wide range of crop and forestry species. There may never be resources sufficient to sequence completely the genomes of all agriculturally important plants, but there is a high degree of similarity among species. Therefore, it is cost-effective to select one representative plant for concerted sequencing of its entire genome.

The model experimental plant *Arabidopsis thaliana* offers the best chance for achieving a completely sequenced plant genome in the foreseeable future. *Arabidopsis* has the smallest known genome of any flowering plant, with 100 million base pairs and little repetition in its DNA. Detailed genetic and physical maps have been developed, and understanding of the plant's biology is growing rapidly. A complete understanding of the *Arabidopsis* genome would offer enormous potential for improving agriculture. In addition to studies on *Arabidopsis*, parallel studies will be required at some level with plants used to grow food and fibre.

Knowledge of a single gene can have broad impact. An example is a recent breakthrough in the identification and sequencing of a tomato gene that confers resistance to a bacterial disease. The sequencing of this gene has given scientists the first unambiguous glimpse of the resistance response of a plant to one of its pathogens. This discovery has opened the way to identification of disease resistance genes in other plants as well as created an avenue for investigating the molecular basis for disease resistance.

Genetics of Rhizosphere Microorganisms

Most farms grow only three or four crops, and some grow the same crops year after year in the same fields. These practices promote soil infestations by pathogens that damage or destroy roots. For most crops, soilborne plant pathogens generally are uncontrolled, except by limited crop rotations and some host plant resistance.

Recent studies show that some soil microorganisms associated with the roots of crops carry and express genes for defence of roots against pathogens. Little is known about these beneficial microorganisms, which represent an enormous untapped genetic resource for improving crop production and efficiency of fertilizer use. Identification and genetic studies of these microorganisms could enable their direct deployment or use of their genes in the development of disease-resistant crops.

Animal Genome Research

The agricultural research community has been quick to take advantage of findings from the Human Genome Project, such as the identification of the gene sequence coding for protein C. Protein C is needed for proper blood coagulation and can be used to treat blood disorders such as hemophilia. Scientists have inserted the protein C gene into swine and developed animals that produce protein C in their milk. The use of swine instead of human blood donations for protein C production would avert the potential contamination of human medications with human serum viruses. Also, production costs would be low because a small number of lactating sows would be able to meet the medical demand for protein C.

An organized, coordinated national research program is in progress to enhance understanding of the structure, function, and regulation of genes in economically important animal species. To achieve maximum effectiveness and cost efficiency, this effort is being conducted in cooperation with other international genome mapping programs. One important focus is on traits controlled by multiple genes.

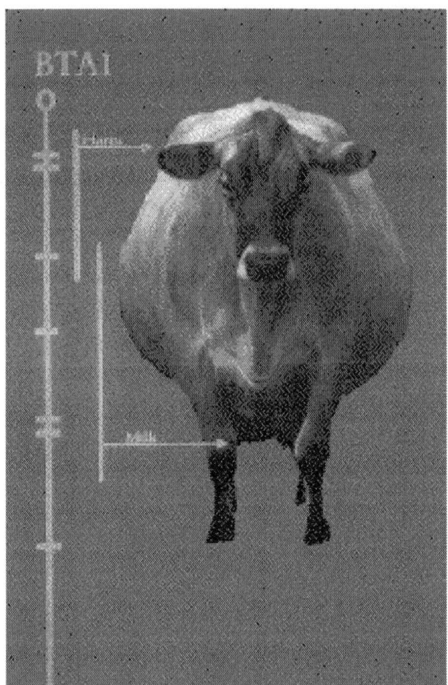

Figure: *Genome mapping research in cattle has identified the chromosomal locations of genes responsible for milk yield and the presence or absence of horns, and it is expected to result in the identification of other genes of economic importance in livestock.*

Most genetic traits that are economically significant are encoded and controlled by quantitative trait loci. In animals, these sets of genes control characteristics such as leanness of meat, meat tenderness, disease resistance, and, in dairy cattle, milk quality and quantity. Recent advances offer the possibility of using molecular markers to improve traits controlled by quantitative trait loci. In some cases, quantitative trait loci can be dissected into individual components, which then become amenable to manipulation. More research is needed to translate the advances made with single-gene traits into advances in understanding of quantitative traits.

Metabolic Studies

Priority: Determine biochemical and genetic control mechanisms of metabolic pathways in animals, plants, and microbes that may lead to products with novel food, pharmaceutical, and industrial uses.

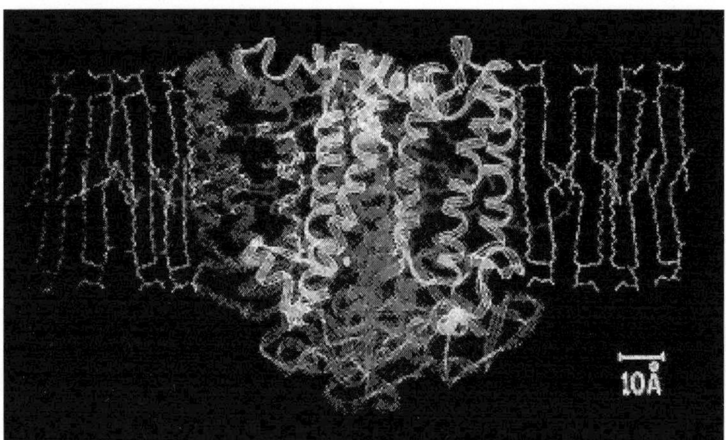

Figure: *This computer graphic rendering of the photosynthetic reaction centre was determined by x-ray analysis. Protein structures are shown as yellow, green and blue ribbons; chlorophyll as a red chain-like structure; the supporting membrane as wavy orange strands; and a single iron atom as a white dot. This and similar structures capture light energy for the formation of biomass, and their efficiency in crop production may be improved by genetic manipulation.*

In the past, new products were developed by exploiting natural materials. Antibiotics were derived from microbes, spices and perfumes from plants, and pharmaceutical agents from plants and other organisms. Plants also have served as renewable resources for energy, building materials, and industrial materials. Today, the tools of biotechnology offer new ways to exploit these biological resources to the maximum benefit of society. The genetic manipulation of plant biochemical pathways can create new products for food and nonfood uses that will expand markets and increase the value of agricultural commodities.

Metabolic studies, which examine the aggregate of the chemical processes carried out by an organism, are a logical focus of Federal support. The ability to add or delete specific genes and change the level of enzyme activity in transgenic plants has given scientists a new way of manipulating and exploring plant metabolism. The development and study of transgenic potatoes, for example, has made it possible to identify the key enzymatic steps in the starch biosynthetic pathway and has led to development of a new variety of potato that has more starch and less water.

Only a fraction of the estimated 100,000 compounds — including lignin, cellulose, terpenoids, and lipids — produced by biochemical pathways in higher plants have been characterized.

The dearth of knowledge is even more extreme for lower plants, such as algae and mosses. There is a pressing need to understand the intricate mechanisms that control various pathways. Research on metabolic pathways and their biochemical and genetic regulation should be expanded, because genes of use to agriculture are likely to be found in diverse organisms. Plants may be the organisms of choice for production of specialty chemicals efficiently and in large volume.

Cell Wall Biomass Products

The cell walls of higher plants are the Earth's most abundant form of biomass. The structure and strength of wood and its industrial products are determined by the nature of the cell walls. Cell walls are the major component of forage crops; the chemical composition of the walls apparently affects feed digestibility. Cell walls provide the mechanical stabilization of the plant body and influence the ultimate size and shape of the plant. Therefore, scientists must understand the biochemistry and physics of cell walls before they can employ biotechnology to modify the architecture of plants.

While progress has been made in understanding plant cell walls, many key questions remain. For example, not a single major enzyme involved in the synthesis of cell wall polysaccharides has been isolated. Information is needed about the organization of individual components within cell walls. And the regulation of the synthesis and assembly of cell wall materials in response to developmental and environmental signals is poorly understood.

New Uses for Plant Materials

Higher plants are very efficient sources of renewable organic materials. In particular, corn starch, which is both a useful raw material and an important human and animal food, is produced in large quantities at about the same cost per pound as is crude petroleum. Many genes involved in starch biosynthesis have been cloned recently and it should be possible to explore the development of new kinds of starches for use in nonfood applications. The potential use of starch as a co-polymer in biodegradable plastics is being explored.

Plant storage lipids (i.e., oils) can be produced in large quantities, and they represent a chemically versatile form of biomass. Enhancement of oil yield and composition in plants such as palm and rapeseed depends on increased understanding of the factors that regulate lipid biosynthesis. Progress has been made in identifying genes and enzymes involved in the early steps of fatty acid and lipid biosynthesis. To date, the most promising approach to manipulating oil yield and composition involves modifying the vast variety of fatty acids found as lipid constituents in various wild species. This is technically possible only if the basic biochemistry is elucidated.

Novel Products from Plants

Higher plants synthesize an enormous spectrum of chemical constituents. Many have known value as drugs, biomaterials, solvents, flavorings, fragrances, or coloring agents. Through development of transgenic plants, the purity of many plant-derived chemicals can be enhanced and the range of chemicals expanded. One example is the pending development of plants that produce hydroxylated fatty acids. These oils can be used in applications ranging from hydraulic fluids to nylon synthesis. Similarly, the recent development of plants that accumulate biodegradable thermoplastics illustrates the potential for producing completely new compounds in plants.

Many natural chemicals are important in plant defence mechanisms or in abiotic stress responses. Some chemicals confer important horticultural qualities to plants. Thorough studies have been conducted on the biosynthesis of certain classes of these compounds, such as flavonoids, cyanogenic glycosides, and certain alkaloids. Other classes of compounds, such as terpenoids, have received relatively little attention. Terpenoids are important for plant growth and can be used as essential oils, resin acids, and pigments. An efficient mechanism must be developed to obtain specific biochemical information about these classes of compounds, to complement the extensive ongoing efforts to identify them and elucidate their structures. Biotechnology offers promise for accelerating the characterization of these substances as well as facilitating their production for commercial use.

Growth and Development

Priority: Extend understanding of the biochemical and molecular basis of growth and development including structural biology of plants and animals. Biotechnology offers opportunities to study a wide variety of genetically regulated growth and development processes. Alteration of these processes can help improve product quality and nutritional and economic benefits. The genetic makeup of plants and animals can be altered by either insertion of new, useful genes or removal of unwanted ones. Discoveries made through biotechnological approaches are changing the way plants and animals are grown, boosting their value to growers, processors, and consumers alike.

This section concentrates on the many opportunities in plant research. The numerous potential benefits range from the development of specialty starch polymers for packaging and oils for use as industrial lubricants to modification of post-harvest traits such as ripening and senescence. The recent development of tomatoes with modified ripening characteristics may lead to extended post-harvest shelf life of fruits, vegetables, and flowers. Information about the biological basis of cold tolerance may result in the development of cold-hardy flowers and ornamentals now grown only in subtropical or tropical climates. The engineering of increased salt tolerance in plants may offset losses due to saline soils.

Plant Flowering

Scientists are gaining their first glimpse into the complex mechanisms of flowering. Most plants use environmental cues to initiate flowering so that individuals of the same species flower synchronously, thus maximizing the chances of successful outcrossing. Scientists have discovered molecules in plants called phytochromes that sense dark and light conditions and send signals that regulate plant activities such as sprouting, growth, and flowering.

There is still much to learn about how flowers develop. Investigations of genes controlling floral meristems — the tissue that forms flowers — promise to help explain the genetic mechanisms that manage the establishment and maintenance of this tissue. Ongoing research also is extending understanding of the physiological processes that take place during the transition from vegetative to reproductive development.

Mechanisms of Plant Fertilization

From apples to zucchini, most foods originate from the fertilized flowers of plants. In tomatoes, for example, several genes have been discovered that act within a flower's anthers, where pollen is produced. Promoters (regions of genes that act as on-off switches) also have been identified. Scientists now are pinpointing the overall molecular mechanisms that cue tomato pollen to fertilize the female part of a tomato flower. Increased understanding of plant fertilization eventually could lead to incremental improvements in the yield and energy efficiency of crop production.

Some plants have genes that render their anthers infertile. This male sterility allows production of higher yielding hybrid seeds, whereby male-sterile plants are fertilized with pollen from specially selected sources. Genetically engineered systems for controlling male fertility are used to produce hybrid seed of important crops such as oilseed rape, corn, and rice. The technology also is applicable to tomato, lettuce, and a wide range of other crop plants. These advances eliminate the need for costly, labour-intensive hand or mechanical removal of anthers in making hybrid crosses and will have a major impact on the billion-dollar hybrid seed industry. In addition, the knowledge gained by studying anther development and fertilization at the molecular and genetic levels will reveal other approaches that can be used to produce novel varieties of hybrid crop plants.

Biological Basis of Plant Form

Conventional breeding often is aimed at achieving certain visual traits. Once the biological basis for these traits is understood, biotechnology could provide the tools to change plant architecture precisely and optimize plant form. For example, changes in leaf form and/or number could maximize photosynthetic capacity and increase production, and changes in root architecture could facilitate and maximize water and mineral capture and uptake. Also, the tools of biotechnology could be exploited to design crops that thrive on soils contaminated with heavy metals, such as cadmium or mercury. Scientists are studying and beginning to redesign the genes responsible for the activity of phytochelatins, natural compounds in plants that bind these metals and detoxify them.

Animal Growth and Development

Research also is needed to explore the genetic basis of animal growth and development. Technologies have been developed for gene cloning, gene transfer, *in vitro* culturing, and sex determination of embryos, and these approaches are being refined for use with animals. Transgenic embryos or offspring have been produced from rabbits, chicken, fish, sheep, swine, and cattle. Genes can be targeted for expression in specific tissues, but the efficiency of gene transfer methods should be improved, and genes important for growth and development must be characterized further.

Environmental Interactions

Priority: Elucidate the molecular basis of interactions of plants and animals with their physical and biological environments, as a basis for improving the organisms' health and wellbeing.

Figure: *Scientists harvesting kernels from genetically engineered barley plants. Plants grown from these kernels will be evaluated for resistance to barley yellow dwarf virus.*

Future gains in agricultural productivity and sustainability depend heavily on the use of biotechnology to improve the health and wellbeing of agriculturally important plants and animals. Research on environmental interactions is emerging rapidly as a powerful means of increasing resistance to stress and disease, with consequent economic and environmental benefits. For example, basic research on the genetic control of biochemical responses to water stress in drought-hardy native plants can suggest how crop plants might be modified genetically for more consistent performance in years of drought. Similar studies are needed to improve the tolerance of U.S. crops to heat stress, winter and frost damage, harm caused by salinity, and other physical stresses.

New crop genotypes also must be resistant to an expanding diversity of pests and diseases. Basic research at the organismal and cellular levels, together with classical breeding techniques, already has paid significant dividends to farmers, consumers, and the environment. Today, U.S. wheat, corn, soybean, barley, and sorghum are grown without fungicides because resistant varieties have been bred. However, these and most other U.S. crops still are subject to major damage from insects and diseases caused by soilborne plant pathogens and insect-vectored viruses, for which there has been no useful source of resistance. As a result, U.S. agriculture continues to depend on pesticides. The survival of high-value horticultural crops depends largely on soil fumigants.

Biotechnology also can help reduce the billions of dollars in agricultural losses caused by animal diseases. There are opportunities to boost the immune competency of animal hosts by genetic manipulation, develop biotechnology-derived regulators of immune function, and identify ways to prevent attachment of disease agents to host cells. Factors affecting the disease-producing capacity of a bacterium or virus can be identified, and strategies can be devised to help the host withstand these pathogens. Through the use of techniques such as enzyme-linked immunosorbent assays, polymerase chain reaction (PCR), and monoclonal antibody-based systems, biotechnology can contribute to major advances in diagnostics.

Plant Pests and Diseases

Virologists have transferred virus genes — such as those for production of virus coat proteins — to plants, thereby conferring resistance to those viruses in otherwise susceptible hosts. Some genes also have been shown to limit virus replication and associated damage.

Replication depends on the regulatory and metabolic processes of the host; these processes must be understood before scientists can engineer virus-resistant plant varieties. Another possible way to increase plant resistance to viral infections would be to interfere with the spread of the virus, but this approach would require an understanding of the biochemical mechanisms of spread and transmission.

All crop plants are susceptible to bacterial diseases, which have been difficult to control through plant breeding. Bactericides are not a complete solution, either, because bacteria quickly evolve resistance to them.

The bacterial disease fire blight is destructive to pears, apples, quince, and some ornamental plants. This disease has plagued fruit producers for over three centuries, but its mechanism of attack only recently has been revealed using the tools of molecular biology. This knowledge is expected to provide ways to combat this disease.

Nearly all agriculturally important fruit, vegetable, and grain varieties acquire fungal diseases that can cost growers billions of dollars annually. Information is accumulating rapidly on the molecular basis of plant responses to infection by these pathogens. In many cases, the difference between resistance and susceptibility is the rate of response; if rapid, then the plant is resistant. Scientists have found that deliberately disarmed fungal pathogens can trigger the necessary response in susceptible (i.e., slow responding) plants before the virulent pathogen arrives. This approach has opened a new avenue in disease control: the use of "pathogen derived" biocontrol agents to induce resistance in otherwise susceptible plants. This technique is only beginning to emerge as a way to control heretofore unmanageable plant diseases.

Control of Insect Pests Preying on Plants

Insect pests cause major crop damage. Examples include *Phylloxera*, which has caused devastation in European and Californian vineyards, and *Psylla*, which has ravaged the apple and pear industry of the eastern United States. In this century, the United States has imported and released approximately 800 natural enemies of insect pests; about 40 percent are providing some level of biological control. However, this approach to insect pest management has its limits. Further progress will require increased knowledge of interactions between pests and their natural enemies, and, if necessary, appropriate genetic modifications.

The chemicals emitted by plants in response to feeding by insect pests serve as an attractant to the natural enemies of some insect pests. Corn leaves attacked by beet armyworm caterpillars release volatile compounds that attract wasps that are parasites of these armyworms. Moreover, these parasitic wasps can recognize and fly toward the source of volatile compounds produced in response to feeding by their caterpillar hosts; the wasps ignore the odours released from leaves damaged mechanically, such as by mowing. Expanded knowledge of the "information molecules" used by beneficial insects to find their prey on crops could lead to development of crops capable of producing stronger signals, to attract higher populations of beneficial insects. To exploit fully this approach to biological control, more research is needed on the biochemistry and molecular biology of insect pests and their natural enemies.

Due to insects' evolving resistance to pesticides, and the removal of many of these chemicals from the market, there are few effective means of controlling sucking insects, such as whiteflies, aphids, and leafhoppers. These same insects are among the most important vectors (carriers) of viruses, rickettsia-like bacteria, and mycoplasma-like organisms responsible for major plant diseases.

New synthetic chemical compounds to be on the market in the next few years primarily target chewing insects. Biological methods to control sucking insects also are advancing at the molecular level. An example is the enzymatic preparation of synthetic pyrethroids.

The salivary glands of aphids, whiteflies, and leafhoppers contain bacterial symbionts that provide essential amino acids to their insect hosts. Only a few of these symbionts have been characterized. Research is needed on the basic biology, molecular biology, and production of these bacterial symbionts. If the bacterial symbionts could be disrupted in some way, either by manipulating the genes of the symbionts or by engineering an antimicrobial agent into a plant, then an innovative control mechanism could be developed.

Beneficial Microorganisms in Plants

Plants selectively aid the growth of specific types of beneficial microorganisms. Some microorganisms, for instance, have been shown to provide growth factors for plants and protect plants against insect attack and infection. Information about the molecular signals and genetic controls of these associations can lead to the use of these microorganisms to improve the performance — and especially the consistency of performance — of important food and fibre crops, and even to enhance the effectiveness of the microbes themselves.

Animal Pests and Diseases

Work is underway to introduce new genes into animals in order to impart resistance to disease and to identify host cell traits that can be used to enhance resistance to disease agents. For instance, a specific receptor on pig intestinal cells is required for *E. coli* to bind to the cell. Strains that do not carry the receptor— or have the receptor deleted through the tools of biotechnology — are less susceptible to the diarrheal disease caused by the bacterium.

Other research on enteric rotavirus infections in pigs has identified specific cell membrane components required for attachment of the rotavirus to an intestinal cell. A complementary molecule can be constructed using biotechnological methods to saturate and block this receptor site. Porcine stress syndrome is an inherited condition in swine that can lead to meat that is of poor quality or unusable. This condition causes significant losses in the swine industry. At the same time, animals susceptible to this genetic disease have highly favourable muscling with low fat content.

A candidate gene for this disease has been identified and research now is needed to develop a DNA probe for the trait, as a means to identify susceptible animals and eliminate them from breeding programs. This research will allow the development of swine with high-quality, low-fat meat but without the trait associated with the disease.

Food Safety

Enhance food safety assurance methodologies, such as rapid tests for identifying chemical and biological contaminants in food and water.

Food-and waterborne illnesses caused by microorganisms pathogenic to humans have been estimated to affect more than 80 million Americans and cost the U.S. economy over $40 billion annually. For example, some 9,000 Americans die each year from food-borne illnesses caused by microorganisms such as *E. coli* in meat and *Salmonella* in poultry. At present, virtually all food inspection is visual. Rapid, accurate, non-invasive tools of biotechnology can help improve the detection and control of food-borne human pathogens as well as chemical contaminants. Federal support is needed in this area to assure that these tools are developed in a timely manner.

DNA Probes

DNA probe kits have been developed for *Salmonella, Listeria, E.coli* 0157:H7, and *Staphylococcus aureus.* Compared to traditional culture-plating methods, these new diagnostic kits offer greater precision, shorter turnaround times, and reduced need for highly trained personnel. DNA diagnostic techniques for Norwalk viruses also are commercially available. Culture techniques have not been successful in detecting these leading causes of gastroenteritis, making the availability of genetically based techniques critical to the detection and identification of the Norwalk viruses.

In addition to detecting food contaminants, DNA probes and other tools of biotechnology can help reduce levels of naturally occurring toxicants in foods. DNA probes can be exploited in research and plant breeding to isolate genes associated with the biosynthesis of major toxicants, facilitate understanding of the genetic regulatory mechanisms for toxicants in plants, and develop lines of plants with reduced levels of toxicants.

Mycotoxins in food are a periodic threat to food safety. DNA probes could help detect the presence of mycotoxin-producing fungi that grow under certain conditions in plant materials such as improperly dried corn and peanuts. DNA probes could also be used to learn more about the sources of fungal contamination in the environment and as a means to develop management strategies under field conditions.

Biosensors

The development of biosensors offers great promise for improving food processing, analysis, and safety assurance. The highly specific actions of biological molecules can be exploited for use in biosensors that can measure the concentration of specific components in complex mixtures. Enzymes, antibodies, and microbial cells can be immobilized on solid surfaces, and the specific reactions they mediate can be detected by various physical and chemical means.

Biosensors are commercially available to detect a variety of sugars, alcohols, esters, peptides, amino acids, cell types, and antibiotics. Development of tailor-made membranes capable of separating molecules based on size, electrical charge, or solubility will accelerate biosensor development as well as the exploitation of biomimetic systems.

Miniaturization and mass production of biosensors could increase their availability and decrease their unit cost. Technologies such as microlithography, ultrathin membranes, and molecular self-assembly have the potential to facilitate the miniaturization of molecular and cellular processes for enrichment, detection, and analysis of chemical and microbiological contaminants in food.

Advances in the semiconductor industry have made it possible to combine chemical and biological components and integrated circuits in miniaturized systems. Biosensors can be inserted directly into food processing streams to obtain on-line, real-time measurements of important food processing parameters.

Miniature biosensors also could be incorporated into food packages to monitor temperature stress, microbial contamination, or remaining shelf life, and to provide a visual indicator to consumers of product state at the time of purchase.

Systematic Study of Food-Borne Microbes

An appropriately designed long-term molecular biology study could characterize and correlate food-associated microbiological isolates in existing culture collections. Food-related isolates from these collections could be used to screen foods for microbiological contamination. Foods identified in this way then would serve as focal points for regulatory agencies and critical systems aimed at improving control over growing, shipping, processing, distribution, and other steps during which these foods may become contaminated.

Opportunities in Environmental Biotechnology

Biotechnology can be used to assess the wellbeing of ecosystems, transform pollutants into benign substances, generate biodegradable materials from renewable sources, and develop environmentally safe manufacturing and disposal processes.

Researchers are just beginning to explore biotechnological approaches to problem solving in many areas of environmental management and quality assurance, such as:

- restoration ecology (which involves reestablishing the nutrient balance and the structure and function of ecosystems);
- diagnostics, epidemiology, and dispersal monitoring related to human disease agents;
- disease, pest, and weed control in agriculture;
- contaminant detection, monitoring, and remediation;
- toxicity screening; and
- conversion of waste to energy.

Environmental biotechnology is not a new field; composting and wastewater treatment technologies are familiar examples of "old" environmental biotechnologies. However, recent developments in molecular biology, ecology, and environmental engineering now offer opportunities to modify organisms so that their basic biological processes are more efficient and can degrade more complex chemicals and higher volumes of waste materials. Notable accomplishments of the "new" environmental biotechnology include the cleanup of water and land areas polluted with petroleum products.

While some success has been achieved, the potential benefits of the new environmental biotechnology are far from fully realized. Advances in this arena are delayed not only by legal and social barriers (which may be formidable) but also by a dearth of basic scientific knowledge about organisms that may be used in biotechnologies and the ecological systems in which these technologies are to be employed. Only new knowledge acquired through Federally supported basic research can provide the foundation for new environmental applications of biotechnology, facilitate the development of these technologies by the commercial sector, and ensure adequate evaluation and safe application of products without blocking innovation with regulatory requirements.

Research in environmental biotechnology has unique international aspects. International cooperation will be needed to help generate new scientific knowledge in this arena, assure U.S. access to the requisite technologies and genetic resources, and establish markets for the resulting U.S. products and processes worldwide.

In addition, environmental biotechnology has tremendous potential for use in developing nations seeking low-cost solutions to environmental problems, such as municipal waste disposal, conversion of agricultural wastes to energy sources, and cleanup of polluted areas. Here, a research program is proposed in one area of environmental biotechnology — bioremediation, an emerging approach to rehabilitating areas fouled by pollutants or otherwise damaged through ecosystem mismanagement.

A novel research framework is described and specific opportunities are identified that offer potential for significant advances. Five research priorities are identified:

- Develop an understanding of the structure of microbial communities and their dynamics in response to normal environmental variation and novel anthropogenic stresses.
- Determine the biochemical mechanisms, including enzymatic pathways, involved in aerobic and particularly anaerobic degradation of pollutants.
- Expand understanding of microbial genetics as a basis for enhancing the capabilities of microorganisms to degrade pollutants.
- As a standard practice, conduct microcosm/mesocosm studies of new bioremediation techniques to determine in a cost-effective manner whether they are likely to work in the field, and establish dedicated sites where long-term field research on bioremediation technologies can be conducted.

- Develop, test, and evaluate innovative biotechnologies, such as biosensors, for monitoring bioremediation *in situ*; models for the biological processes at work in bioremediation; and reliable, uniform methods for assessing the efficacy of bioremediation technologies.

Bioremediation is addressed as one example of an environmental biotechnology. Because the knowledge required for bioremediation is similar to that needed for the development of many other environmental biotechnologies, the research approach described here is likely to have wide application.

Bioremediation: an Overview

The United States has a large number of identified polluted areas, including land, fresh water, and marine sites that, by law, must be cleaned up. Estimates for the cleanup of Federal lands alone may be $450 billion. The extent of contaminated non-Federal agricultural acreage, mining areas, industrial sites, and aquifers and other water bodies is unknown, but the magnitude of the problem is undoubtedly large and clean-up expenses could be astronomical. It has been estimated that cleanup of both Federal and non-Federal lands could cost $1.7 trillion using conventional approaches, which would produce noxious waste by-products and thereby impose additional clean-up or environmental costs.

Due to its comparatively low cost and generally benign environmental impact, bioremediation offers an attractive alternative and/or supplement to more conventional clean-up technologies.

Bioremediation has been successful at many sites contaminated with petroleum products. However, it is not always the technology of choice because efficacy and the rate of degradation at any particular site cannot be predicted reliably. Improved predictive and process validation capabilities would help stimulate wider use of this technology. Research also could lead to development of biotechnologies to remediate areas contaminated by metals, pesticides, radioactive elements, other toxic materials, and mixed wastes.

These types of studies could be especially productive at this time. Recent developments in biology have provided new tools and approaches for monitoring the environment and engineering organisms with the capacity to degrade environmental pollutants.

These developments have created unprecedented opportunities for significant advances. Indeed, bioremediation is expected to become an industry with annual sales of more than $500 million by the year 2000. The United States is among several nations developing bioremediation technologies.

Maintaining and enhancing the U.S. position in this arena will require continued investment in the generation of new knowledge needed for the development of new technologies. Investment in bioremediation research has the dual benefits of solving important environmental problems while stimulating the growth of the U.S. bioremediation industry.

Proposed Research Approach

Key to the success of the investigations proposed in this chapter is the overall structure of the research program. The complex environmental milieu in which bioremediation and other environmental technologies will be employed demands an holistic research strategy; the traditional, piecemeal approach will not be adequate. A research program designed using an holistic approach would

- Form a continuum: Bioremediation research generally is conducted at one of three scales: laboratory, pilot scale, or field trial. To help ensure that results achieved at the first two scales can be translated to the field, the research program should be conceived as a continuum, with investigators working at each scale involved throughout the research conceptualization and planning process. The aim is to translate research findings from the laboratory into viable technologies for remediation in the field.
- Be multidisciplinary: The biological and physical complexity of the field environment, where research findings ultimately will be tested and applied, demands a multidisciplinary research team composed of microbiologists, ecologists, engineers, hydrologists, and other specialists.
- Encompass temporal and spatial variation: A major impediment to achieving the full potential of bioremediation is the lack of access to field sites for iterative research on basic mechanisms, technology development, and process verification. Field research sites should be identified that accommodate long-term experimentation on a large spatial scale.
- Address the complexity of "real world" situations: Many polluted sites contain mixtures of wastes of such chemical complexity that a suite of biochemical processes is required for degradation.

Waste mixtures may contain sanitary (household) waste plus hazardous organic or inorganic materials. Hazardous waste is defined legally as toxic, corrosive, reactive, or ignitable. "Mixed waste" generated by Federal weapons and research complexes is a combination of radioactive and hazardous waste.

Therefore, research focusing on the degradation of mixtures of wastes by groups of organisms is likely to be most realistic and efficacious.

Focus on Microorganisms

Different types of organisms can be bioremediation agents. For example, the use of plants to concentrate pollutants (phytoremediation) is an emerging research area. This chapter, however, examines bioremediation by microorganisms, the most biochemically diverse and least understood group of organisms on Earth.

Microorganisms (primarily bacteria and fungi) are nature's original recyclers. Their capability to transform natural and synthetic chemicals into sources of energy and raw materials for their own growth suggests that expensive chemical or physical remediation processes might be replaced or supplemented with biological processes that are lower in cost and more environmentally benign.

Microorganisms therefore represent a promising, albeit largely untapped resource for new environmental biotechnologies. Research continues to verify the bioremediation potential of microorganisms.

For example, a recent addition to the growing list of bacteria that can sequester or reduce metals is *Geobacter metallireducens*, which removes uranium, a radioactive waste, from drainage waters in mining operations and from contaminated groundwaters. Even dead microbial cells can be useful in bioremediation technologies. These discoveries suggest that further exploration of microbial diversity is likely to lead to the discovery of many more organisms with unique properties useful in bioremediation.

A tiny fraction of the microbial diversity of the Earth has been identified, and an even smaller fraction has been examined for its biodegradation potential. Research aimed at characterizing this diversity likely would lead to the discovery of novel mechanisms for biodegradation of pollutants. Once the biochemical and ecological nature of microbial biotransformations of pollutants is understood, new bioremediation technologies based on these mechanisms will be possible.

Fundamental Research Areas

Expanded research is needed on the basic ecology of microorganisms and interactions among microbial community members. In nature, microorganisms seldom exist or act as single species; instead, they act collectively as consortia. Research examining bioremediation of polychlorinated biphenyls (PCBs) in the Hudson River, for example, revealed that both anaerobic and aerobic bacteria, acting together, were responsible for degrading the pollutant. (Anaerobes grow in the absence of oxygen, while aerobes require it.)

Additional research of this type will provide a framework for understanding how microbial communities respond to various environmental stresses; how to accelerate *in situ* bioremediation by native microbial communities; and whether the introduction of engineered microbes with enhanced bioremediation potential can survive and function within established communities and help remediate the site.

Such studies are likely to provide corollary insights into aspects of microbial biochemistry important for bioremediation as well as the roles of microorganisms in biogeochemical cycling.

Physiology/Biochemistry Research

Priority: Determine the biochemical mechanisms, including enzymatic pathways, involved in aerobic and particularly anaerobic degradation of pollutants. Much of the information relevant to bioremediation has come from studies of the genetics and physiology of aerobic bacteria. As a result, the best-known biochemical processes related to bioremediation are oxygen dependent. This characteristic limits their effectiveness in many polluted underground and underwater sites with minimal or no oxygen.

Both biological and physical strategies for improving the supply of oxygen in such sites have been proposed, and research on aerobic organisms must be continued. However, long-term success in dealing with a wide array of polluted sites with little or no oxygen will require information that can be obtained only through increased research on the genetics, physiology, and biochemistry of anaerobic organisms.

In nature, biodegradation in sites with little or no oxygen is mediated by anaerobic and microaerophilic microorganisms. Because of the difficulty of isolating and culturing such organisms in the laboratory, their metabolic diversity and their potential use in environmental biotechnology only recently have been appreciated.

These technical obstacles are being overcome with improved cultivation methods, new technologies for identification of microorganisms, and new methods for studying their metabolism *in situ*.

New knowledge about anaerobic microorganisms has expanded opportunities for exploiting their metabolic diversity in bioremediation. Since the discovery in 1982 that some anaerobes can dehalogenate carbon compounds, microbes with this capability have been detected in a wide variety of anaerobic environments, such as the Hudson River sediments. However, to exploit anaerobes for bioremediation, more knowledge is needed about the biology of diverse anaerobic microbes, including how they respond to fluctuating oxygen levels.

Because the concentration of oxygen in soil and water environments is often highly variable (in time and space), biodegradation in nature probably is mediated by both aerobic and anaerobic microorganisms. This phenomenon can be exploited through the development of anaerobic-aerobic technologies for controlled bioremediation. In other words, the environment can be modified either spatially or temporally to allow the development of various types of biodegradative microbial communities, thereby fostering the degradation of pollutants vulnerable to different agents. Anaerobic-aerobic processes can be developed to exploit the full range of microbial metabolic activity for cleanup of environments contaminated with multiple pollutants.

Genetic Research

Priority: Expand understanding of microbial genetics as a basis for enhancing the capabilities of microorganisms to degrade pollutants.

The successful use of microorganisms in bioremediation depends on the development of a basic understanding of the genetics of a broad spectrum of microorganisms, many of them not yet isolated or studied in any detail. Microorganisms adapted to degrade specific pollutants have been found among populations growing naturally at polluted sites. However, the genetic mechanisms underlying specific adaptations are poorly understood.

In the past, researchers have been unable to conduct genetic studies on these hard-to-culture organisms, but recent developments in molecular biology now make it possible to isolate and study genes of almost any organism. Scientists now can analyse genes that govern a wide variety of metabolic processes, including the degradation of environmental pollutants. Such genetic analyses provide information about mechanisms underlying the operation and evolution of degradation pathways. While the isolation and characterization of novel microorganisms and genes are important in developing bioremediation strategies, knowledge about the general features of the microbial genome also are needed. Bacteria exhibit a high degree of genetic plasticity, or changeability. Analysis of complete microbial genomes will reveal the nature of this plasticity and suggest how genetic engineering can be used to modify organisms to impart the characteristics needed for bioremediation.

New knowledge about the diversity of microorganisms and the organization of their genes also will help explain how environmental factors influence the expression of genes and the regulation of microbial metabolism. Research has revealed that genes expressed when one compound is present can play a role in the metabolism of a second compound.

For example, some bacteria were found to degrade highly toxic trichloroethylene (TCE) when the less toxic compound toluene was present. Mutant organisms also have been isolated that can degrade TCE in the absence of toluene due to genetic changes that cause the TCE-degrading gene to be "turned on" continuously.

However, bacteria capable of TCE degradation may lack other traits — such as tolerance for heavy metals, salts, and acid soils — needed for their use in bioremediation. Genes for these traits could be added through genetic engineering, thereby increasing the degradation efficiency of native microorganisms.

Recombinant microorganisms with expanded degradation capabilities have been developed recently. Researchers in several countries studied a number of microorganisms, each of which degraded a restricted range of pollutants, and characterized the genes involved. Investigators then combined genes from different species into one strain of bacteria that can degrade multiple types of pollutants.(7) Similar applications of modern genetic techniques should make it possible to tailor bacteria for bioremediation of sites contaminated with specific combinations of toxic compounds.

From the Laboratory to The Field: The Research Continuum

Priority: As a standard practice, conduct microcosm/mesocosm studies of new bioremediation techniques to determine in a cost-effective manner whether they are likely to work in the field, and establish dedicated sites where long-term field research on bioremediation technologies can be conducted.

Priority: Develop, test, and evaluate innovative biotechnologies, such as biosensors, for monitoring bioremediation *in situ*; models for the biological processes at work in bioremediation; and reliable, uniform methods for assessing the efficacy of bioremediation technologies.

Microcosm/Mesocosm Studies

Once organisms or groups of organisms with bioremediation potential have been identified through laboratory screening or genetically engineered microcosm/mesocosm studies — scale-up studies conducted in large bioreactors or on small, protected areas of land or water — can indicate whether the organisms are likely to perform as desired at field scales. This intermediate step permits control of the pollutant concentration, the numbers of degradative agents, and the physiochemical environment.

More specifically, the dissipation of the pollutant can be quantified and the breakdown products measured and identified using labels and tracers that would be too costly to employ in the field. Changes in the degradative capacity and ecological adaptation of the degrading microorganism can be monitored, and environmental conditions that favour bioremediation can be assessed. If non-native organisms are being evaluated, then their ability to survive and compete with the native microbes can be determined. In addition, potential adverse effects of the bioremediation strategies studied can be detected and mitigated.

Many bioremediation applications benefit from testing and scale-up from the laboratory to the mesocosm level. An example is the use of bioreactors for the immobilization of metals from water or the treatment of industrial and municipal waste, before the techniques are used in the field.

When degradation is examined under intentionally varied environmental conditions, the results can suggest whether biodegradation is likely to occur in the field at an acceptable rate with or without oxygen and nutrient augmentation. It is useful to employ systems of various sizes, ranging from small laboratory microcosms, such as soil columns and biofilms, to soil lysimeters (large, instrumented *in situ* blocks of soil isolated by physical barriers).

While the results obtained from laboratory bioreactors cannot be extrapolated directly to predict field performance, research at the microcosm/mesocosm scale can reveal genetic and biochemical changes in organisms exposed to the target pollutants. This approach may save considerable time and resources by suggesting whether a remediation system has any chance to succeed in the field.

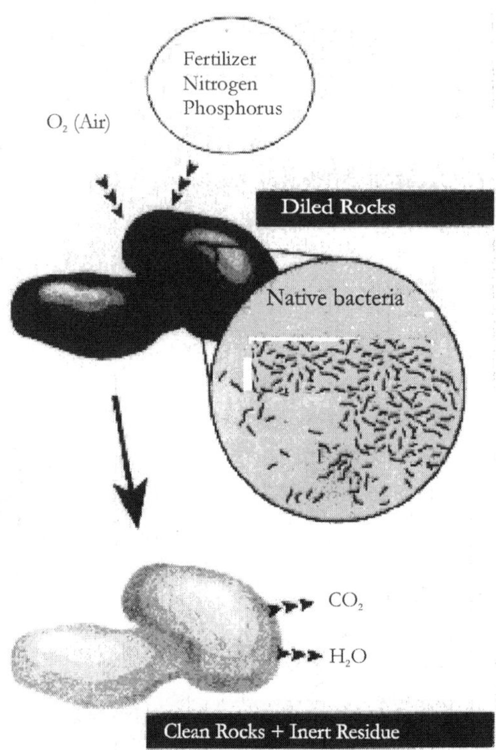

Figure: *In this example of bioremediation, natural biodegradation of oil was accelerated by the addition of fertilizer which balanced the amounts of nitrogen and phosphorus with the available carbon. When present in optimal amounts, the three elements can be fully utilized by microorganisms to form carbon dioxide and water from oil.*

Field Studies

The successful development and application of bioremediation technologies depends on field-based research to verify the efficacy of planned approaches under natural conditions. Field research is a complex undertaking, in part because any number of problems can arise. For example:

- an immunological probe may not be effective due to binding of the substrate to the soil or soil organic matter;
- an engineered microorganism may not be competitive with the indigenous population, or it may not survive at certain times of the year or under specific moisture regimes;
- biocontrol and environmental monitoring technologies are unlikely to be effective in all regions, due to differences in plant

communities, soils, climate, disease vectors, land management practices, and economic policies;

- geological and soil conditions may preclude delivery of bioremediative microorganisms to polluted sites; or
- supposedly "clean" sites may continue to leach contaminants as chemical and physical conditions change over time.

A further challenge in field evaluations of bioremediation technologies is the need for data collection over long periods of time and at various spatial scales. It can be difficult, if not impossible, to find suitable field sites. A recent report from a workshop sponsored by the American Academy of Microbiology (AAM) concluded that:

> *A major problem in the development of bioremediation technology is the lack of field sites that are well-characterized with respect to contaminants, geohydrology, and geochemistry; such sites are urgently needed for understanding the natural events that are taking place and also for the transfer of technology developed in the laboratory to field conditions. An integrated interdisciplinary approach is essential for the application and verification of bioremediation, and this can only be achieved under environmental conditions. Although some sites may already exist, their openness and flexibility of use is unlikely to support more than a few efforts in the bioremediation community.*

In response to this conclusion, the AAM held a follow-up workshop to address strategies and mechanisms for field research. The report from the second workshop recommends that field sites have specific characteristics and defines the goals of the interdisciplinary research to be conducted at such sites. Although some Federal programs are beginning to address the critical need for field experimentation identified by the two AAM reports, there are no long-term, interdisciplinary bioremediation research programs conducted at dedicated field sites.

To accelerate the discovery and development of effective bioremediation technologies, several long-term research sites should be established. Coordinated, interdisciplinary research should be conducted to explore the bioremediation-related characteristics of these sites. Then, multidisciplinary groups of researchers should define cooperatively the basic questions to be answered, the scope of the work to be accomplished, and the methods of validation of field results.

Integral to the research conducted at these field sites would be assessment of the efficacy of the biotechnologies employed. As new bioremediation data are collected and verified, new field experiments could be planned and discoveries transferred to the private sector for development and commercialization.

A coordinated program of long-term research at field laboratories could facilitate the development of risk assessment baseline information, including bioavailability data; promote the transfer of all findings to industry and other potential user groups; generate quantitative information on mechanisms of passive remediation as a bioremediation alternative; foster the development of biosensors for process characterization and validation; and, perhaps most importantly, establish a database that could be used to predict conditions under which bioremediation can be achieved.

Efficacy Testing Methods and Data Evaluation

One barrier to the acceptance and use of bioremediation technologies is the lack of established methods for determining efficacy in the field. Evaluation of bioremediation in the field is difficult due to the complexities of the natural environment, the competing abiotic processes, and the vast temporal and spatial scales involved.

The AAM has identified the lack of adequate performance standards as a key factor undermining user confidence in bioremediation. The recent growth of the bioremediation industry has intensified the problem of determining the effectiveness of a large number of competing processes.

The development of standardized evaluation protocols would be facilitated by research comparing various methods for obtaining evidence. Possible measures include numbers and activities of bacteria, changes in inorganic carbon concentration, or the extent of transformation of pollutants into forms that can be biodegraded more easily. Although innovation might be inhibited by premature setting of standards or overregulation of the fledgling bioremediation industry, methods for standardized efficacy testing and data evaluation should be developed as an integral part of the industry's growth. Ideally, to ensure objectivity, a government agency not directly involved in site cleanup or regulation would help develop standardized testing methods and materials.

As efficacy studies are carried out, a large volume of historical data will be generated. The quality and comparability of such data must be evaluated and controlled carefully.

Biosensors: One Promising Assessment Technology

Currently, *in situ* degradation processes cannot be measured or validated directly; researchers must rely on tracers and gas generation to assess bioremediation processes. Increased investment in biosensor research could lead, over the long term, to improved tools for efficacy assessment. The development of biosensors promises to revolutionize the way pollutants are detected and monitored in the environment.

Unlike standard methods, which rely on analytical chemistry in measuring the total concentration of a pollutant, biosensors can detect the fraction available to microorganisms. Biosensors also have the advantage of being nondestructive and located on-line, meaning that samples do not have to be removed and transported to a laboratory for analysis. Biosensors may utilize either whole bacterial cells or specific molecules (e.g., enzymes or biomimetics) as a detection system. Combinations of biosensors in arrays can be exploited to deal with a diversity of toxicants and pollutants.

One type of biosensor involves linking a gene such as the mercury resistance gene (*mer*) or the toluene degradation (*tol*) gene to genes that code for bioluminescence within living bacterial cells. The biosensor cells can signal that extremely low levels of inorganic mercury or toluene are present in contaminated waters and soils by emitting visible light, which can be measured with fibre-optic fluorometers.

A second type of biosensor employs molecular detectors, which consist of enzymes, nucleic acids, antibodies, or other "reporter" molecules attached to synthetic membranes. Antibodies specific for an environmental contaminant can be coupled to changes in fluorescence to increase sensitivity of detection. Fluorescent or enzyme-linked immunoassays have been derived for a variety of contaminants, including pesticides and PCBs. With sustained long-term research, molecular arrays could be constructed on synthetic membranes and other matrices that would allow the simultaneous detection of a range of contaminants in a variety of environmental substrates.

Modelling

Enzyme redesign is an emerging area of research that holds great promise for improving the bioremediation potential of microorganisms. Enzyme models can be created and manipulated by supercomputers to change primary enzyme structure and then predict resulting changes in enzymatic specificity and activity. Changes that would enhance bioremediation activity then can be engineered into bacteria, and those bacteria tested for a desired result in microcosms and mesocosms.

In the absence of reasonable parameters for the biological processes involved in bioremediation, models also can be developed to indicate whether physical factors also are at work.

Models could be used to distinguish the important variables from those with only minor effects and to extrapolate results for one geographic region to another, based on predicted patterns of interactions between physical and biological factors.

Opportunities in Manufacturing/Bioprocessing Upstream Processing

Priority: Investigate methods to enhance the efficiency and expand the utility of upstream processing technology.

Upstream processing encompasses any technology that leads to the synthesis of a product as well as the fundamental science and engineering needed to understand product formation. Specific areas of opportunity for Federal research include biocatalysis, metabolic engineering, biomass conversion, bioreactor design and cell culturing techniques, and transgenic animals.

Biocatalysis

Nature's catalysts — enzymes — function highly selectively to accelerate the rate of chemical reactions in biological systems. Enzymes also can limit reactivity toward a particular substrate and limit chemical conversion to a single desired product. For many substrates, enzyme-catalyzed reactions exhibit rates that are many orders of magnitude faster than those of uncatalyzed reactions.

In chemical synthesis and other processes of industrial importance, the ability to mimic the selectivity and rate-enhancing properties of enzymes would be of substantial benefit in reducing economic and environmental costs. In some cases, this approach might provide a commercial means of synthesizing a material where only much less desirable methods were available previously. At present, about 50 enzymes are used in industry.

Most of these industrial uses involve breaking down large molecules into simpler ones. Examples include the proteases in laundry detergents, and the hydrolases used in starch processing. The commercial benefits of enzymology would be enhanced if technologists could expand capabilities for creating complex molecules from simpler ones or transforming existing chemical structures into more active compounds.

Production of sweetness-enhanced corn syrup by the enzyme glucose isomerase is one noteworthy example of how a chemical transformation can be carried out on the industrial scale. Synthesis routes that can distinguish between right-and left-handed chemical structures are being pursued for pharmaceutical, food, and agricultural applications. Many industrial firms are exploring the unique capabilities of enzymes to carry out such chemical transformations.

Federally sponsored research can facilitate the development and application of new enzymes of industrial importance. Many research efforts are under way, and solid progress has been made using protein engineering, chemical modification, and recombinant monoclonal antibody technology. More recently, recombinant DNA technology has been used to generate catalysts that exhibit the useful characteristics of naturally occurring enzymes but have been designed either to catalyze chemical reactions not found in nature or to alter biological reactions to broaden their specificity and expand their potential for practical application.

Progress in this field will accelerate with an improved understanding of relationships among protein structure, function, and energetics. More efficient methods are being developed to obtain protein structure using X-ray crystallography, nuclear magnetic resonance (NMR), and neutron scattering techniques, coupled with robust algorithms to treat the data and display the results using computer graphics. The use of computer graphics to visualize the dynamics of proteins also will enhance understanding of the mechanisms of enzyme action. Progress in obtaining structural data in real time ultimately will help modelers develop algorithms to predict the dynamics of enzymes. Research also should be supported on the thermodynamics and kinetics of prototypical enzyme-catalyzed reactions of interest to the bulk and commodity chemical markets; examples include the reactions involved in corn processing and the action of laundry detergents.

Metabolic Engineering

Metabolic engineering is the use of recombinant DNA technology to enhance the activities of a cell by manipulating its metabolic pathways. To exploit this approach, scientists must develop an improved understanding of cell metabolism. Experimental and mathematical techniques are needed for quantifying the effects of altered uses of raw materials by cells. Techniques such as electrophoresis can be used to distinguish and quantify cellular proteins produced in response to altered metabolism. In addition, non-invasive techniques such as near-infrared spectroscopy and NMR can quantify intracellular concentrations of raw materials. Also of interest is research on thermodynamics and kinetics of prototypical microbial and microalgal systems.

The first applications of metabolic engineering have incorporated genes for pathway enzymes and transport proteins. In this manner, for example, the bacterium *E. coli* has been induced to produce ethanol. But amplified expression of non-native proteins can affect host cell metabolism — sometimes interfering with the engineering objective — because raw material resources are reallocated. The effects may include a slowing of cell growth and possibly cell death. Often, the expression of small amounts of non-native proteins increases the level of stress proteins. This problem is ubiquitous, occurring in simple organisms such as *E. coli* and in complex cells, such as yeast cells, human cells, the chinese hamster ovary cell (used commercially for protein synthesis), and an insect cell line that is a possible source of proteins for making AIDS vaccines. Federal support for research in this area can promote efforts to overcome or mitigate these undesirable cellular responses.

Biomass Conversion

Biomass includes organic polymeric material — such as lignin, starches, celluloses, and oils — produced by biological processes. Plants and algae are the main sources of biomass, generating many billions of tons annually through photosynthesis. Other sources of biomass derived from human activities include food processing wastes, waste paper, and municipal solid wastes.

The Federal Government should expand its support for research that will facilitate development of commercial products from this abundant raw material. Biomass is a potential source of energy resources such as methane and ethanol, commodity chemicals, animal feed, and specialty products (e.g., flavors, fragrances, pigments). Biomass is an attractive alternative to petroleum-based sources because it not only is a low-cost substrate for production, but also is environmentally friendly; it is renewable and may reduce emissions during processing and improve the biodegradability of end products.

Biotechnology is especially relevant to the commercial use of biomass because it employs natural processes uniquely suited for creating specific products. An example of such a process is fermentation, which converts glucose sugars derived from biomass into commodity chemicals. The components of plant biomass, and products that can be derived from them, include

- cellulose (a mostly crystalline polymer of glucose molecules), which can be hydrolyzed (i.e., broken down) to form glucose, which in turn can be bioconverted into a variety of products, such as ethanol;
- hemicellulose (another heterogeneous, amorphous sugar polymer), which can be hydrolyzed to make sugars such as xylose and glucose;
- lignin (a complex aromatic polymer), which can serve as a chemical feedstock or be burned to provide heat and electricity;
- starch, used by the corn wet milling industry and fuel alcohol industry as the basis for large, commercial-scale bioprocesses; and
- oils. Production of diesel fuel from oil seeds is probably the most well known industrial application of a "renewable" oil; the hydroxy fatty acids from oil seeds can be used in a wide range of products, including cosmetics, waxes, nylons, plastics, and coatings.

Fuels and chemicals can be produced not only from oil seed crops (e.g., soybeans, canola, and crambe), but also from unicellular algae and waste fats from animals. The substantial fractions of oils in these resources can be transformed through a chemical process that changes the large molecules in fats and oils into smaller molecules.

Bioreactor Design and Cell Culturing Techniques

If the great variety of potentially commercial bioprocesses are to be developed and applied, then bioreactors must be designed in which the environment can be controlled precisely to maximize process efficiency. Federal support for research in this area can improve the efficiency of bioprocessing and thereby improve prospects for its wide application.

The design of a bioreactor requires a basic understanding of both chemical reactor design and cell biology. First, designers must understand the effects of reaction rates and stoichiometry, mass transfer, heat transfer, and turbulence and mixing on product distribution, reactor productivity and size, and operational characteristics. These phenomena need to be expressed in accurate but tractable models that can be used for design and optimization calculations.

In order to develop effective cell culturing techniques, designers also must have basic knowledge of cellular functions and protein chemistry. They should understand the molecular, genetic, and metabolic processes involved in the growth of cells and the expression of cellular products; and structure/function relationships in the use of proteins for biochemical conversions.

Transgenic Animals

Transgenic animals can be employed either as biofactories for the production of commercial products or as living models for the study of human diseases and evaluation of pharmaceuticals. The use of transgenic animals for these purposes can be more economical or, in the case of human disease and drug models, more realistic than are conventional alternatives. Federally supported research can improve the efficiency of the processes involved and identify new opportunities.

Transgenic animals are being developed for a wide variety of applications. Transgenic cows are used for production of lactoferrin, a naturally occurring milk protein used in products such as baby formula. Several companies plan to market pharmaceuticals from the milk of sheep and goats and the blood of pigs. Anti-clotting factors used to treat cardiovascular diseases have been produced in rabbits, goats, and mice.

Research also is under way to develop transgenic animals — mice in particular— that carry genes associated with human diseases. Such animals might enable researchers to identify specific disease genes or produce human antibodies.

Downstream Processing

Priority: Develop capabilities to recover and purify products from dilute bioprocess streams and develop predictive models to facilitate the design of downstream separations.

Downstream processing includes the cost-effective separation and purification of bioproducts, as well as biorefining. Often, downstream processing is the most expensive phase of producing a substance of biological origin, especially for products with stringent regulatory requirements. A key reason for this high cost is the complex and dilute nature of the aqueous solutions in which bioproducts generally are produced; an inverse relationship has been demonstrated between the price of biological products and the strength of the concentrations from which they must be isolated. This is the case for bulk and commodity chemicals, including ethanol from biological sources.

The high cost of downstream processing means there is significant potential for savings from improved processes. That factor, coupled with the broad applicability of downstream processing research, makes this area a prime target for Federal investment. For regulated bioproducts, such as pharmaceuticals, removal of trace impurities is the expensive step in the purification.

Separation and Purification

The separation and purification of materials produced in a bioreactor is a critical part of a manufacturing operation. The biological products involved range from high-value-added substances used as pharmaceutical agents (e.g., insulin) to lower-cost products including commodity chemicals (e.g., ethanol).

High-value bioproducts are usually fragile molecules, such as proteins or peptides, that require highly specialized and mild processing conditions and may need to be separated from a complex mixture of molecules, including cell debris. This combination of factors makes separation difficult. At present, most separation schemes are scaled-up laboratory procedures; research is needed to improve their performance.

Biological methods hold promise for improving separations. For example, a cell could be modified genetically to suppress production of undesirable by-products and alter the cell wall so it would be permeable only to the desired product. It is also possible to genetically engineer a product so that it passes through the cell wall, has an affinity for certain separation matrices, or is joined to another molecule with desirable separation characteristics. In addition to improving separations, biological approaches also reduce energy expenses, a major part of the cost of separating biological products.

Biological processes one day may offer economical alternatives to current, petrochemically based methods for manufacturing organic acids and alcohols. However, before these bioprocesses can become commercially viable, nontraditional, lower-cost separation methods need to be developed. Research is under way to develop extracting solvents, resins (separation media), and sorbents that are more selective and have a higher capacity than do current materials.

Reversible extraction systems are needed that respond to changes in temperature, pressure, or acidity. Combinations of conventional separation methods and biological methods are being explored to reduce product inhibition, which often occurs in fermentations that produce alcohols and solvents.

In addition, mathematical models of separation steps need to be developed to help reconcile regulatory requirements with basic process conditions during early stages of process development, and to meet the demands of a competitive business environment. Modelling enables scale-up considerations to be estimated very quickly, a capability needed in order to commercialize a bioprocess in an industry where being the "first to market" is a critical element of success.

Biorefining

Biorefining involves the use of microbes in mineral processing systems. Biorefining is environmentally friendly and in some cases enables the recovery of minerals and use of resources that otherwise would not be possible. For example, a significant amount of copper is biorefined from slag heaps. Research now under way or planned addresses the use of microorganisms to bioleach oxide and sulfide ores, and to concentrate metals.

Current research on bioleaching of oxide and sulfide ores addresses treatment of manganese, nickel, cobalt, and precious metal ores. The objective is to identify metabolic pathways in the microorganisms responsible for metal solubilization, and to improve their survival rates and stability. Increased understanding of metabolic pathways will open the door to the manipulation of parameters such as kinetics and metal selectivity, with the aim of enhancing mineral recovery.

An intriguing potential use of biotechnology is *in situ* bioleaching of ore deposits or waste piles. Research is needed to develop a mechanistic understanding of the bioleaching process and to identify environmental and process factors affecting biosystem performance. In addition, bioprocess monitoring schemes and cell-free leaching systems should be developed. Cell-free or cell component systems, in which cellular components rather than live organisms are used for a specific reaction, eliminate the need to provide an environment conducive to survival of a microorganism. Such approaches also tend to enhance control over the reaction, due to the specificity of the cell component.

Biosorption and metal recovery from dilute aqueous solutions is an emerging field of interest, from both a resource conservation standpoint and an environmental remediation standpoint. Microbes, algae, cell wall material, proteins, and other types of biomass have been investigated for use in this application, but economical methods for selective metal recovery have yet to be developed. The Federal Government should support research in this area. Inexpensive substrates for protein immobilization should be developed, and biomolecules with high metal-binding capacity identified.

Use of microorganisms to remove fine particles from aqueous process streams is also of interest. Laboratory-scale studies have shown that certain microorganisms can cause flocculation of fine mineral suspensions, while others can function as flotation collectors or depressants. Extensive research will be needed before the use of extracellular colloidal molecules (such as polysaccharides and exopolymers released by live biomass) can become a viable alternative to current chemical-based systems.

Process Monitoring and Control

Priority: Develop methods for monitoring and control of commercial bioprocessing including reliable real-time sensors.

Process monitoring and control requires knowledge of the current state of every step in a bioprocess and the ability to control the process for optimal results. The most efficient use of upstream and downstream processing depends on knowledge of the state of the system and on control algorithms that can optimize the process and maintain it. This is another area where Federal support for generic research can have significant payoffs for multiple industries.

Control and optimization is challenging due to several characteristics of bioprocesses, including their high degree of nonlinearity (meaning that nonlinear differential equations are required for mathematical modelling), especially in batch and fed-batch bioreactors; and their potential for instability when they involve high-yield mutant or recombinant organisms. There are technical obstacles as well, including the unavailability of reliable on-line real-time sensors and realistic models that capture the complexities of biological systems by identifying and quantifying rate-controlling steps in reactions.

Optimization and control methods should be robust, adaptive, and suited to nonlinear processes. Important components of these systems include biosensors and bioelectronics/bionetworks. Because an optimal process often is operating at the limits of process variables, process control can be very difficult; consequently, design methods should consider simultaneously both the economics and the controllability of the process. The affordability of some production processes, particularly for lower-value products, depends on the development of highly accurate sensors and models and the capability to reproduce bioreactor operation precisely.

Biosensors

Sensors are invaluable in the design and operation of automated and environmentally benign manufacturing processes, and in detection, monitoring, and control of food additives, food safety factors, and bioremediation technologies. Biosensors are attractive because they can harness the inherent specificity of many biochemical processes to allow for accurate and highly specific detection of organic and inorganic species in real time. Biotechnology can be used to design, construct, probe, and alter sensitive and highly specific biosensors and incorporate them into transducing and reporting systems.

A biosensor encompasses three major components: a biological component, an interface, and a transducing element. The biological component (e.g., enzyme, immunoprotein, nucleic acids, whole cell) interacts specifically with the substance to be detected. The interface component (e.g., polymeric thick or thin film, chemically modified surface) links the biological component with the transducer. The transducer converts the biochemical interaction into a quantifiable electrical or optical signal.

An example of a biosensor is the plasmon sensor, which consists of a monolayer of immunoproteins immobilized on the surface of a thin film of gold. The analyte — whatever compound binds to the protein employed — attaches to the immunoprotein and changes the properties of surface electromagnetic waves that can be excited in the gold film. This change, which signifies the presence of the analyte, can be detected optically.

A vast store of fundamental information is available on the biological components used in biosensors. The key challenge now is to improve the manufacturing process so that reproducible, dependable, low-cost biosensors can be mass produced. Federal support can hasten progress in this area and foster the use of biosensors in a broad range of applications.

There are a number of technical obstacles. Poor adhesion of polymer films to the transducer, and hence poor reliability, is a major problem. The film must be compatible with the biological component and also adhere to the transducer surface. Another problem is that insufficient knowledge of methods and effects of immobilizing biological components in/on polymeric films has led to wide variation in sensor response, complicating the manufacture of commercial biosensors. Non-specific binding must be minimized, because the specificity of analyte detection is degraded when other molecules compete for binding sites.

There are also operational difficulties that preclude the use of biosensors in bioprocessing environments. For example, many sterilization techniques (such as heat) either modify or denature biological molecules. There are also problems with miniaturization, sensitivity, dynamic range, and noise reduction related to specific biosensor fabrication strategies. Finally, there is the question of quality control — determining quickly and nondestructively whether the biosensor is operating with prescribed precision and accuracy.

Bioelectronics and Bionetworks

Bioelectronics is an emerging technology that employs biological molecules instead of inorganic materials in conventional integrated-circuit technology or in applications involving unconventional architectures, such as optical processors. The driving force for this research is the possibility of constructing devices on the molecular level and thereby achieving extremely high densities of data storage sites and nano-sized computers.

Biological systems are capable of storing information on the molecular level and processing information along pathways defined on a molecular level. Although biological processes function more slowly than do conventional solid-state devices, this penalty is more then offset by a huge increase in the density of operating units. Various biomimetic or biologically based materials, such as the protein bacteriorhodopsin, are being evaluated for use in bioelectronics.

In living things, data processing is achieved by arrays of neurons. Although the operation of single neurons is well understood, the operation of biological neural networks remains largely unexplored. Recent achievements in the culturing of a monolayer of neurons on a micro-electrode array promise to provide some insight into the operation of neuron arrays. Although neuron devices — bionetworks — would not operate on a molecular scale, they have the potential to form the basis for new computer architectures, including parallel processors.

Appropriate Federal investments to address key research questions could help make the United States a world leader in this promising new arena. The major technical obstacle to development of practical bioelectronic devices lies in determining how to preserve and control the properties of the active state when the bioactive species is immobilized in a artificial membrane. Research to explore the fundamental process of biological self-assembly, a capability inherent in biological molecules such as lipids, DNA, and proteins, may be useful in resolving some of the organizational problems. The pattern inherent in a bi-lipid layer or an alkane thiol monolayer ("smart" organic compounds) on a metal surface can be used as a template for further organization of biomolecules.

Biomaterials Processing

Priority: Expand the development of novel biomaterials, such as biomimetics and replacement tissues, through new tissue engineering and chemical synthesis methods.

Up to this point, this chapter has focused on generic processing technologies used to make a variety of products. This section addresses the processing of one particularly promising category of products — biomaterials. The tools of biotechnology can be employed to endow materials with properties not achievable using more conventional means. This research area deserves steady Federal support because of the important applications for biomaterials, which include biological substitutes for human tissues.

Biomolecular Materials

Due to their diversity, versatility, and unusual combinations of properties, biomolecular materials offer promise for application in virtually all sectors of the economy, including defence, energy, agriculture, health, and environmental technology. Examples of biomolecular materials include silk obtained from spiders, and the ceramics in sea shells.

In addition to their direct use as natural cellular products or modified derivatives, biomolecular materials serve another very important purpose by demonstrating how nature has optimized their physical properties. Research is needed to clarify how higher-order structure is achieved and how it serves to determine macromolecular function in such a variety of forms.

With continued advances in modern biology, molecular genetics, and protein engineering, and with rapid improvements in physicochemical characterization, novel biomolecular materials can be designed and tailor-made to meet specific needs.

This, in turn, would expand the possibilities for practical applications of these materials. Standard chemical synthesis has inherent limitations, including the production of compounds with unwanted impurities and by-products.

By contrast, biomolecular synthesis— manufacturing methods based on biological processes — allows precise control, thereby reducing levels of impurities and by-products. For instance, when a cell produces peptide polymers, control of the amino acid sequence is assured by the fidelity of RNA and DNA replicative mechanisms.

Thanks to recombinant DNA technology and continuing advances in molecular biology, that same control over uniformity of composition, length, and sequence now is available to the scientist seeking to synthesize and express natural or tailor-made genes for peptide polymers.

By inserting appropriate DNA sequences into microorganisms, algae, and higher plants, investigators have learned how to induce the cells to synthesize polymers (such as polylysine) that they do not ordinarily produce, in quantities sufficient for further study and eventual application. Research is needed to define the classes and specific types of polymers that can be produced in this manner. Federal support could play a key role in advancing research in this area.

Tissue Engineering

Tissue engineering — a term coined in 1987 referring to the development of biological substitutes to restore, maintain, or improve human tissue function— employs the tools of biotechnology and materials science as well as engineering concepts to explore structure-function relationships in mammalian tissues. This emerging technology could provide for substantial savings in health care costs and major improvements in the quality and length of life for patients with tissue loss or organ failure.

Advances in the study of tissue growth and regeneration, at both the cellular and tissue levels, set the stage for practical application of tissue engineering. The culturing of cells in two-dimensional monolayers enabled the study of cellular processes and opened the door to genetic manipulation. Scientists and engineers have begun to view cell culturing as a three-dimensional process, in which external forces on the cells not only may influence cellular products, but also may reawaken the cellular differentiation process. In order to develop living tissue equivalents, it will be important to understand how the cellular environment affects the differentiation process as well as interactions between the engineered tissue and the host.

The first success with differentiated cells came with engineered human skin, now in clinical trials. Scientists also are beginning to explore the potential to grow many tissues in culture. Using stromal cells from human tissue, researchers are developing blood vessels, bone, cartilage, nerve, oral mucosa, bone marrow, liver, and pancreatic cells. Federal support can hasten progress in development of these materials.

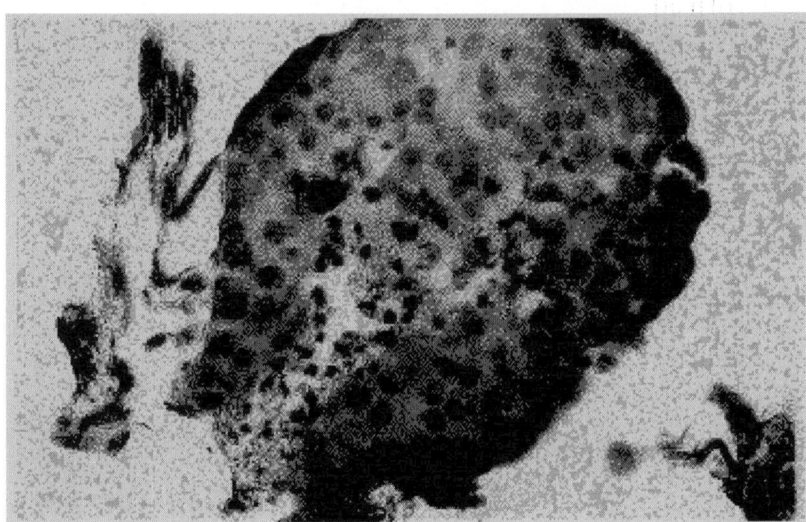

Figure: *Rotating tissue culture vessels, which simulate some aspects of microgravity, allow human cells to grow and assemble into 3-dimensional tissues.*

Encapsulated cell therapy is an example of a technique under development by industry that employs biomaterials in the treatment of certain serious, chronic diseases. The goal of this approach is to replace cells within the body that have been destroyed by disease in order to augment circulating or local levels of the deficient molecules. Targets for replacement include insulin-producing cells in diabetics and dopamine-secreting cells in patients with Parkinson's disease.

An encapsulated cell implant consists of cells that secrete the desired hormones, enzymes, or neurotransmitters, enclosed within a polymer capsule and implanted into a specific site within the host. Animal studies have shown that the functional activity of secreting cells can be maintained in vivo. The capsule wall is designed to allow passage of small molecules (i.e., glucose, other nutrients, therapeutic molecules) but prevents or retards the passage of large molecules, such as elements of the immune system. Studies suggest that the transplanted cells are protected from destruction and perhaps even recognition by the host's immune system, allowing the use of unmatched or even genetically altered tissue without systemic immunosuppression.

Thus, the use of an encapsulating membrane may overcome two of the difficulties that prevent widespread tissue transplantation into humans: the limited supply of donor human tissue, and the toxic effects of immunosuppressive drugs required to prevent rejection of unencapsulated transplants. This is an important research area where expanded Federal investment could yield significant dividends.

Opportunities in Marine Biotechnology and Aquaculture new and Improved Products from the Seas

Priority: Develop a fundamental understanding of the genetic, nutritional, and environmental factors that control the production of primary and secondary metabolites in marine organisms, as a basis for developing new and improved products.

Priority: Identify bioactive compounds and determine their mechanisms of action and natural function, to provide models for new lines of selectively active materials for application in medicine and the chemical industry. Most major groups of living organisms are primarily or exclusively marine. Tropical marine environments harbour an especially wide diversity of animals and plants. Many marine organisms are sessile and must employ sophisticated methods to compete for a place to anchor. This characteristic is reflected in part by a metabolism that produces enormously diverse bioactive products — many with no terrestrial counterparts.

Recent research has uncovered unicellular and multicellular microorganisms that are unique to the marine world. Indeed, "... it seems clear that marine bacteria are emerging as a significant chemical resource" and the amounts and characteristics of chemicals they produce.

Federal support for research in this area is essential, both to answer the fundamental questions and to assure that the resulting knowledge is translated into sustainable technologies. While it will be vital to cultivate marine microorganisms that produce novel products, the alternative approach of transferring genes of interest into non-marine microorganisms also should be investigated.

For example, the capability to produce a marine polysaccharide — a complex molecule that could be useful as a food additive or a water-resistant adhesive — could be transferred to an easily grown bacterium (e.g., *E. coli* or *Bacillus subtilis*). This approach might be more effective in some cases than would cultivating the marine organism or recreating artificially the long and complicated production pathway for the polysaccharide.

Pharmaceuticals

Many bioactive substances from the marine environment already have been isolated and characterized, several with great promise for the treatment of human diseases. The compound manoalide from a Pacific sponge, for example, has spawned more than 300 chemical analogs, with a significant number of these going on to clinical trials as anti-inflammatory agents. To promote sustainable technology development, the Federal Government should support research leading to new methods for discovery. One approach would be to focus on learning about natural functions, regulation, and production of substances generated by marine organisms, in order to identify potentially important agents. Refined test systems should be developed in order to identify selective agents produced by marine organisms. Rapidly developing assay technology can facilitate exploration of the bioactivity of newly discovered compounds. These assay methods, which employ specific receptors for known physiological agents, require only minute amounts of a test substance and can be automated. (Traditional chemical tests require considerable amounts of test material and laborious measurement processes.) The new tools will make it feasible to test newly discovered compounds rapidly by the hundreds, for a wide spectrum of biological activities.

To date, exploitation of natural agents from the sea has been hindered by problems with limited or sporadic distribution and production. Much more research must be conducted to determine what seasonal factors and life cycle or reproduction states are linked with natural production of an agent. Factors influencing production may include diet, physical and chemical conditions, distribution by phylogenetic affiliation, geographic location, water depth, or associations with symbiotic microorganisms. Knowledge of these factors will be important in developing methods for producing selected metabolites, either from whole organisms or *in vitro* from cell or tissue cultures of plants and animals. Many of these compounds are very large and complex molecules, requiring very elaborate biochemical processes; as a result, it can be difficult to synthesize them or clone all the genes for standard production through fermentation.

Enzymes

Enzymes produced by marine bacteria are important in biotechnology due to their range of unusual properties. Some are salt-resistant, a characteristic that is often advantageous in industrial processes. The extracellular proteases are of particular importance and can be used in detergents and industrial cleaning applications, such as in cleaning reverse-osmosis membranes. *Vibrio* species have been found to produce a variety of extracellular proteases. *Vibrio alginolyticus* produces six proteases, including an unusual detergent-resistant, alkaline serine exoprotease. This marine bacterium also produces collagenase, an enzyme with a variety of industrial and commercial applications, including the dispersion of cells in tissue culture studies. Other research has demonstrated the presence in algae of unique haloperoxidases (enzymes catalyzing the incorporation of halogen into metabolites). These enzymes could become valuable products, because halogenation is an important process in the chemical industry. Japanese researchers have developed methods to induce a marine alga to produce large amounts of the enzyme superoxide dismutase, which is used in enormous quantities for a range of medical, cosmetic, and food applications. An unusual group of marine microorganisms from which enzymes have been isolated are the hyperthermophilic archaea (previously called archaebacteria), which can grow at temperatures over 100° C and therefore require enzyme systems that are stable at high temperatures. Archaea typically are found in extreme environments, such as hot springs, animal guts, hydrothermal vents, sewage sludge digesters, and hypersaline habitats, including the Great Salt Lake.

Thermostable enzymes offer distinct advantages, many still to be discovered, in research and industrial processes. Thermostable DNA-modifying enzymes, such as polymerases, ligases, and restriction endonucleases, already have important research and industrial applications. Hot springs in Yellowstone National Park provided the first archaeon from which thermostable DNA polymerases were isolated. These novel enzymes (the *Taq*® polymerases) became the basis for the polymerase chain reaction (PCR), a useful technique for studying genetic material. In 1989, thermostable DNA polymerase was designated Molecule of the Year by Science magazine. Comparable enzymes continue to be discovered.

Most enzymes involved in the primary metabolic pathways of thermophilic bacteria and archaea are dramatically more thermostable than are their counterparts living at moderate temperatures. Expanded study of enzymes from thermophilic marine microorganisms will contribute to the understanding of mechanisms of enzyme thermostability and should enable the identification of enzymes suitable for industrial applications as well as modification of enzymes to enhance thermostability.

Biomolecular Materials

Recent research has demonstrated that marine biochemical processes can be exploited to produce new biomaterials. For example, a corporation in Chicago is commercializing a new class of biodegradable polymers modelled on natural substances that form the organic matrices of mollusk shells. Equally exciting are the mechanisms used by marine diatoms, coccolithophorids, mollusks, and other marine invertebrates to generate elaborate mineralized structures on a nanometer scale (less than a billionth of a meter in size). Nanometer-scale structures can have unusual and useful properties.

Research that will enhance understanding and allow engineering of the processes for creating these bioceramics promises to revolutionize the manufacturing of medical implants, automotive parts, electronic devices, protective coatings, and other novel products. Biomaterials also hold promise for counteracting biofouling, which long has been recognized as an extensive and costly problem. Bacterial biofilms form slime layers that increase drag on moving ships, interfere with transfer on heat exchangers, block pipelines, and contribute to corrosion on metal surfaces. Bacterial and microalgal colonization of surfaces is accompanied by settlement of invertebrate larvae and algal spores, eventually leading to "hard fouling" and the need for costly cleaning.

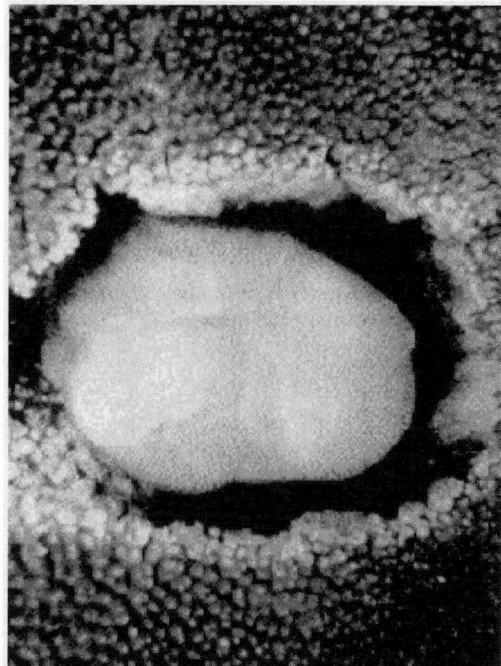

Figure 8: Sessile marine organisms often produce potent defensive chemicals, which, when released into seawater, create a protective zone around the organism. Through strong biological activity these chemicals deter predators and protect the producer from encroaching neighbours, fouling larvae or settling microorganisms.

The most effective anti-fouling coatings have utilized toxic chemicals, such as copper and organotins. There is an urgent need for nontoxic biofouling control strategies, due to heightened recognition of the impact that toxic coatings can have on the environment. Research is needed on the attachment mechanisms of marine organisms and the natural products they employ to prevent fouling of their own surfaces.

Molecular approaches to characterizing biofilm structure and development offer considerable potential for finding novel biofouling prevention strategies. It is now possible to determine the genes and pathways involved in regulation and synthesis of bacterial adhesive polymers. Considerable progress has been made in understanding the nature and expression of surface polymers produced by microorganisms such as the nitrogen-fixing *Rhizobium* species and the opportunistic pathogen *Pseudomonas aeruginosa*. Similar approaches can been applied to marine biofilm bacteria, to find the genetic determinants of adhesive production and the environmental factors that regulate synthesis.

Molecular biology techniques also can be applied to determine the basis of natural antifouling mechanisms. Many marine plants and animals remain free of attached bacteria, either because they produce repelling compounds or because their surface structure neutralizes bacterial adhesives. A product generated by the seagrass *Zostra marina* (eelgrass), for example, is an effective agent for preventing fouling by bacteria, algal spores, and a variety of hard-fouling barnacles and tube worms. Molecular characterization of natural fouling resistance could provide new strategies for fouling control. Potential applications include prevention of fouling in industrial pipelines or heat exchangers, improved design of trickling filters or aquaculture circulation systems, and control of biofilm infections of medical implants and prosthetic devices.

Biomonitors

Marine organisms can provide the basis for development of biosensors, bio-indicators, and diagnostic devices for medicine, aquaculture, and environmental monitoring. One type of biosensor employs the enzymes responsible for bioluminescence. The *lux* genes, which encode these enzymes, have been cloned from marine bacteria such as *Vibrio fischeri* and transferred successfully to a variety of plants and other bacteria. The *lux* genes typically are inserted into a gene sequence, or operon, that is functional only when stimulated by a defined environmental feature. The enzymes responsible for toluene degradation, for example, are synthesized only in the presence of toluene. When *lux* genes are inserted into a toluene operon, the engineered bacterium glows yellow-green in the presence of toluene. This genetically engineered system "reports" that biodegradation of a specific chemical, in this case toluene, is proceeding.

Another type of biomonitor that holds great promise is the gene probe, which can be used to identify organisms that pose health hazards or may be useful in research. Specific gene probes can be employed, for example, to detect human pathogens in seafood and recreational waters; fish pathogens in aquaculture systems; microorganisms capable of mediating desired chemical transformations (e.g., toxic chemical degradation, CO2 assimilation, metal reduction); and specific fish stocks in fish migration and recruitment studies.

Biopesticides

Natural marine products have the potential to replace chemical pesticides and other agents used to maximize crop yields and growth. Continued Federal support for R&D in this area is likely to result in useful natural pesticides that would provide greater specificity and fewer harmful side effects than do conventional synthetic agents. Current U.S. expenditures for all pesticides amount to $47 billion annually; by the year 2000, biopesticides from marine and other sources are expected to capture an estimated 10 percent of this market.

An example of a marine biopesticide in use today is Padan*TM*, which was developed from a bait worm's toxin known to ancient Japanese fishermen. This natural pesticide has demonstrated activity against larvae of the rice stem borer, the rice plant skipper, and the citrus leaf miner, among other pests. More recently, scientists in Montana discovered novel compounds in marine algae and marine sponges containing symbiotic microorganisms. These compounds promoted growth and stimulated germination and increased root and coleoptile lengths in test plants. Several sponge and nudibranch species produce terpenes, a broad class of aromatic compounds used in solvents and perfumes and known to deter feeding by fish. Extracts derived from these same sponge and nudibranch species also demonstrated powerful insecticidal activity against two species, grasshoppers and the tobacco hornworm.

Biomass for Energy Production

Approximately 40 percent of all primary energy production, or photosynthesis, occurs in the seas. In this process, oceanic plants (phytoplankton, seaweeds, seagrasses) take up carbon dioxide (CO_2) and, with light energy from the sun, convert it into organic carbon (primarily sugars) and oxygen. The oceans contain 50 times as much carbon dioxide as does the atmosphere, and it is estimated that primary production incorporates 35 gigatons (1 gigaton = 1 x 10*15* grams) of carbon into marine biomass annually.

This abundant source of fuel for energy production has not been tapped commercially because it is not competitive with soybean meal and other easily harvested, traditional sources of biomass, and also because, regardless of the source, biomass is not competitive with other types of fuels.

The Federal Government should continue to support research on the use of biotechnology to enhance biomass production and utility. At least three general approaches are being explored.

First, the enzyme that captures CO_2 for photosynthesis — ribulose bisphosphate carboxylase\oxygenase or "RUBISCO" — is relatively inefficient, so supercomputers are being used to verify structural information, and the enzyme is being redesigned to optimize its function. Second, the chemical composition of biomass can be altered to make it more suitable for particular applications. For example, marine microalgae are being genetically engineered to boost their lipid content, with the aim of providing a source of alternative fuels that is more economical than are conventional sources. Third, biotechnology is being used to convert biomass to ethanol and other alternative forms of energy and chemical feedstocks.

New and Improved Processes from the Seas

Priority: Develop bioremediation strategies for application in the world's coastal oceans, where multiple uses — including wastewater disposal, recreation, fishing, and aquaculture — demand prevention and remediation of pollution; and develop bioprocessing strategies for improving sustainable industrial processes.

Bioremediation

Bioremediation shows great promise for addressing problems in marine environments and in aquaculture. These problems include catastrophic spills of oil in harbours and shipping lanes and around oil platforms; movement of toxic chemicals from land, through estuaries, into the coastal oceans; disposal of sewage sludge, bilge waste, and chemical process wastes; reclamation of minerals, such as manganese; and management of aquaculture and seafood processing waste.

The full potential for marine organisms and processes to contribute new waste treatment and site remediation technologies cannot be realized without enhanced understanding of the unique conditions in marine environments. For example, oxidation-reduction (redox) states can fluctuate in coastal and estuarine sediments. The impact of changing redox conditions on biodegradation of environmental contaminants must be understood before waste management and remediation strategies and predictive models can be developed for contaminated sediments.

Bioprocessing

The emerging discipline of bioprocess engineering involves the application of biological science in manufacturing, to produce products such as biopharmaceuticals and natural bioactive agents. Bioprocess engineering requires an understanding of the biological system employed (such as a marine organism), isolation and purification of a product, and translation of the product into a stable, efficacious, and convenient form. An emerging area of interest is the potential of marine bacteria and fungi to produce unusual chemical structures with no parallels in terrestrial organisms. Small-scale studies have begun to indicate the richness of marine microorganisms as sources for novel lead structures.

Aquaculture

Priority: Use the tools of modern biotechnology to improve the health, reproduction, development, growth, and overall wellbeing of cultivated aquatic organisms; and promote the interdisciplinary development of environmentally sensitive, sustainable systems that will enable significant commercialization of aquaculture.

Aquaculture, which long has been practiced in Asia and is increasingly popular in the United States, Europe, and South America, will benefit tremendously from the use of new molecular tools and processes. With worldwide seafood demand projected to increase 70 percent in the next 35 years, and harvests from capture fisheries stable or declining, aquaculture will have to produce seven times as much seafood as it generates now to supply global demand by 2025. The use of modern biotechnology to intervene in the rearing process and enhance production of aquatic species holds great potential not only to meet this demand, but also to improve U.S. competitiveness in aquaculture.

The U.S. aquaculture industry has grown rapidly in recent years. Farm gate receipts exceeded $800 million in 1992 — a fourfold increase since the early 1980s.

The growth and international competitiveness of the U.S. aquaculture industry will be determined by the size of the resource investment in research and technology development. This investment should be made through a partnership of Federal and state agencies and the private sector. The Federal role is to provide leadership in supporting research to advance knowledge in important research areas and to facilitate the transfer of promising results and technologies to the private sector.

The major research issues in aquaculture are similar to those for other agricultural sectors, but the knowledge base for aquaculture is comparatively meager. Development of this knowledge is a particular challenge due to the diversity of cultured aquatic species and the systems for their production. Federal support for biotechnology research in this area will expand the knowledge base and yield significant dividends. The application of biotechnology promises significant benefits to both producers and consumers of aquacultural products. The use of genetically enhanced organisms may improve production efficiency through improvements in growth rates, food conversion, disease resistance, and product quality and composition.

The application of biotechnology to aquaculture also may help conserve wild species and genetic resources and provide unique models for biomedical research.

Enhancing Reproduction and Early Development

Biotechnology can be applied to enhance reproduction and early development of cultivated aquatic organisms. The resulting benefits could include year-round production of gametes and fry of economically valuable species and creation of new markets for specialized, genetically improved broodstock. Similarly, biotechnology may provide techniques for improving the reproductive success and survival of endangered species, thereby helping to preserve the diversity of life on Earth.

As a first step, research should be directed toward improving basic understanding of environmental, hormonal, biochemical, and genetic control of reproduction. More specifically, scientists must identify and understand the mechanisms of expression of genes involved in reproduction and development, improve technologies for cryo-preserving gametes and embryos, improve delivery systems for administration of natural and synthetic hormones, and enhance understanding of the pharmacokinetics of uptake and release of administered hormones.

Figure: *Experimental facility for study of hormonal control of salmon development. Location Washington State*

Improving Health and Well-Being

Biotechnology offers substantial opportunities to improve the health and wellbeing of cultivated aquatic organisms. More than 50 diseases affect fish and shellfish cultured in the United States, causing losses of tens of millions of dollars annually.

Biotechnology not only can improve the survival, growth, vigour, and wellbeing of cultivated stocks, but also can reduce disease transfer between cultivated and wild stocks. New products and market opportunities can be developed related to aquatic animal health and wellbeing.

The tools of molecular biology can provide a basic understanding of host immunity, resistance, and susceptibility to diseases and associated pathogens by furnishing information about life cycles and mechanisms of pathogenesis, antibiotic resistance, and disease transmission. Improved technologies must be developed for detecting and diagnosing pathogens and diseases and for enhancing the genetic basis of disease resistance, thereby reducing the need for antibiotics and other drugs.

Potential products resulting from this research include gene therapy techniques; broodstock free of pathogens; safe, effective prophylactic agents, including immune modulators, antigens, and vaccines; safe, effective therapeutic agents; and improved systems for administering prophylactic and therapeutic agents.

Improving Quality and Value

The Federal Government has a responsibility to help ensure the safety and quality of food supplies, and biotechnology can and should be an invaluable tool in carrying out this mandate. Biotechnology can be employed to assess and improve the safety, freshness, colour, flavour, texture, taste, nutritional characteristics, and shelf life of aquacultural food products. In addition, practical technologies can be developed to detect and assay toxins, contaminants, and residues in seafood, and to reduce or eliminate contaminants. There are also opportunities to apply biotechnology in improving seafood processing. Research should be conducted to develop and improve technologies for all these applications.

Conserving Genetic Resources

The preservation and enhancement of biodiversity in natural systems is an important Federal priority. Therefore, the Federal Government should encourage and support programs to maintain and enhance biodiversity in aquatic systems through cultivation and stocking of aquatic species. Biotechnology can be employed in two ways to conserve genetic resources of aquatic species.

Figure: *Harvesting hybrid striped bass from experimental ponds used in Sea Grant research.*

First, the tools of biotechnology can be used to identify and characterize important aquatic germplasm, including endangered species. Genomes of aquatic species can be analysed and characterized, and quantitative trait loci identified. Second, biotechnology can be applied to improve understanding of the molecular basis of gene regulation and expression as well as sex determination and thereby improve methods for defining species, stocks, and populations. Approaches include developing marker-assisted selection technologies, improving precision and efficiency of transgenic techniques, and improving technologies for the cryo-preservation of gametes and embryos. Ultimately, stocking certain areas with selected, cultivated species and strains could help maintain biodiversity in natural aquatic ecosystems.

Enhancing Biomedical Models

Aquaculture has important purposes other than food production. Because they often adapt to extreme environments, marine organisms can provide unique models for research on biological and physiological processes. Studies of the developmental, cellular, and molecular aspects of marine organisms as model systems will provide insights into the basis of disease mechanisms and pathogenesis in humans. By contrast, use of mammalian organisms as a basis for the development of some types of human disease models may be neither feasible nor cost effective.

Progress in this arena will require that sophisticated molecular biology technologies be adapted to marine organisms, in order to enhance understanding of their biological processes. For example, approaches for gene transfer into eggs have been developed for many terrestrial organisms, but not for most marine species. This technology is needed for analyses of gene regulatory systems and gene expression. In addition, methods need to be developed for culturing tissues from marine organisms. Cultured cell lines will provide opportunities for gene transfer and gene expression studies and enhance the usefulness of marine species as biomedical research models. This is an important research area deserving of Federal support.

Understanding And Conserving The Seas

Priority: Improve understanding of microbial physiology, genetics, biochemistry, and ecology in order to provide model systems for research and production systems for commerce, and to contribute to understanding and conservation of the seas.

Scientists have a powerful new array of sampling devices and measuring instruments that will accelerate greatly the acquisition of knowledge about ocean resources and foster their wise use. These technologies include manned deep-sea submersibles, remotely operated vehicles, geosynchronous satellites, sophisticated acoustic measuring devices, pressure-retaining deep-sea samplers, geographic information systems, real-time flow cytometry, PCR and biomonitoring techniques, computerized databases, and other forms of information exchange and analysis. These tools should be exploited to accelerate the discovery of unknown marine microorganisms and to expand understanding of known varieties. Federal support for this research is essential, because only then will sufficient information be acquired to assure that practical applications will result. As new life forms and processes become known, and as understanding of them grows, marine biotechnology will make significant contributions to the nation's social and economic wellbeing.

Identifying Organisms and Their Niches

Biodiversity

Nowhere in the biosphere is biological diversity greater than in the seas, and the extent of this diversity becomes increasingly evident as scientists investigate new environments. In the 1980s, for example, giant tube worms (*Riftia*) were recovered from areas adjacent to deep-ocean thermal vents, and novel mussels (*Bathymodiolus*) that farm methanotrophic bacteria on their gill tissue were discovered around methane seeps in the Gulf of Mexico. Most newly described species have been and will continue to be microorganisms, although it is clear that new marine plants and animals also await discovery. Less than 1 percent of the extant bacteria— marine and terrestrial — have been isolated and described. The marine environment represents a particularly fertile source for new bacteria, as evidenced by the recent discovery of unusual "cold water" archaea 100-500 meters deep in the oceans. These archaea comprise a high percentage of the total bacterial ribosomal RNA present in seawater samples, yet they have not been isolated in pure culture and described. Pursuit of this research may provide new understanding of oceanic processes and, once these archaea are cultivated, perhaps a novel source of products. Countless other marine bacteria also have yet to be cultured; advances in microbial culture technology will enable scientists to isolate, describe, and possibly exploit these organisms.

Viruses are another newly appreciated element of marine biological diversity. Electron microscopy and PCR technology have revealed abundant numbers of viral particles in seawater samples. Specific viruses that infect species of marine phytoplankton have been cultured from coastal as well as open ocean sites. This discovery is exciting for at least two reasons. Viral infection of higher forms of marine life almost certainly affects global ocean processes, such as photosynthesis. Marine viruses also will provide new materials for development of genetic and biotechnological tools that can be used to study and manipulate marine organisms. For example, marine viruses could be used to genetically engineer higher forms of marine life.

Marine Ecology

As a basis for developing new applications for marine products and processes, analysing global climate change, and improving fisheries management, scientists must build a fundamental understanding of marine organisms and their specific adaptations to and interactions with their environment. Recent developments in molecular techniques are rapidly expanding capabilities for research on marine ecology. Three research areas seem to be especially fertile and deserving of Federal support.

As a first step, the new molecular tools should be applied to gain insight into the basic molecular and cellular processes by which marine organisms adapt to extreme environments. Examples of basic research that may lead to commercial applications are gene sequencing projects and the ongoing study of special glycoproteins that inhibit ice crystal growth in the tissues of Antarctic fish. These substances, along with gene sequence information for key marine species, may prove useful in industrial and medical preservation processes.

Second, a thorough understanding of marine ecological systems must be developed in order to specify the "normal" baseline level of function and to monitor and predict potential changes and perturbations of systems due to physical, chemical, or biological impacts. The development of predictive models for analysing potential global climate changes depends on the acquisition of fundamental information on molecular regulatory mechanisms of photosynthesis in the oceans.

Third, research on marine ecology can be conducted to benefit fisheries management. The tools of biotechnology can be used to determine the effect of natural and anthropogenic perturbations on the size of commercial stocks, to delineate fishery stocks and management units of living resources, to determine predator-prey relationships, and to restore habitats essential to robust fisheries.

Defining the Impact of the Seas on the Global Environment

Powerful tools are being developed to elucidate the many biogeochemical cycles that determine the fate of all the life-supporting elements on Earth. Scientists are beginning to understand and manipulate the molecular genetics and biology of esoteric metabolic pathways associated with the carbon, sulfur, phosphorous, iron, and other biogeochemical cycles. For example, based on the hypothesis that iron controls photosynthesis in the oceans, immunological probes were used to show that addition of iron to open ocean water off the Galapagos Islands significantly increased energy production.

Marine biotechnology will be useful in assessing the role of the oceans in affecting climate change and the global carbon cycle. Molecular techniques can facilitate and enhance the measurements of CO_2 concentrations and total CO_2 inventories being developed for global ocean models of carbon cycling.

There is compelling evidence that the exchange of dissolved and particulate materials between the continental shelf and its boundaries is a significant factor affecting the flux of CO_2 and biogenic elements within the global ocean. Several Federal agencies plan to collaborate on related research, including marine biotechnology applications.

Infrastructure Needs

Physical Infrastructure Needs

Funding for facilities has been very low across all agencies. In fiscal year (FY) 1993, only $53.7 million was invested across the Federal Government. Deterioration of existing facilities and lack of funds for new facilities for biotechnology research threaten to limit progress. The broader problem of the deterioration of academic research facilities has caught the attention of the Administration and the Congress, resulting in a substantially increased appropriation for academic infrastructure in the FY 1995 budget of the National Science Foundation.

The total scope of this problem cannot be dealt with adequately by any individual agency. While specialized needs must continue to be supported to the extent possible through existing programs, a coordinated interagency effort to upgrade research facilities is needed. There are a number of urgent needs specific to biotechnology research:

- Synchrotron facilities and neutron facilities for structural biology research need to be upgraded and constructed.
- Researchers need to have access to containment facilities for recombinant DNA research at the P-3 and P-4 safety levels. This need is especially acute in some areas of the country and for academic institutions with relatively small, but nonetheless important, commitments to biotechnology research.
- Plant growth chambers and greenhouses need to be constructed, repaired, and replaced.

- Animal facilities need to be constructed, renovated, and maintained to meet animal care guidelines and regulations emanating from the Animal Welfare Act.
- Large, specialized containment facilities, analogous to greenhouses, for bioremediation research, need to be provided to permit controlled growth of microorganisms as well as contamination studies on scales that mimic those of natural sites. In addition, one or several examples of large contaminated natural environments should be selected as field sites for bioremediation to permit studies on appropriate physical and temporal scales over long time periods.

Instrumentation

Equipment and instrumentation are critical not only to support the research of high-caliber scientists, but also to help attract and train the next generation of biotechnology researchers and technologists. Typically, the equipment and instrumentation required for biotechnology are very sophisticated and therefore expensive to acquire, operate, and maintain. One cost-effective solution to this problem is development of instrumentation centres available to researchers from government, academia, and all industry sectors. Many different types of instruments are required:

- Specialized chemical hoods, large fermentors, and autoclaves are needed to deal with large volumes of contaminated materials in environmental research.
- New, highly sensitive and portable analytical instruments are needed for *in situ* bioremediation monitoring.
- There are needs for capital-intensive instrumentation, such as X-ray crystallography and nuclear magnetic resonance equipment, computer capabilities (particularly the coming generation of parallel processors), and advanced mass spectrometers for use in manufacturing/bioprocessing research.
- There are needs for specialized equipment for the collection, maintenance, husbandry, and cultivation of marine organisms; specialized fermentation and process-recovery equipment capable of withstanding the corrosive effects of sea water, for the production of marine products; and improved sampling equipment and techniques to replace indiscriminate, mass collections of organisms in the field.
- Equipment and associated instrumentation is needed for culturing of, and bioprocessing with, marine organisms and

organisms from extreme environments, including hyperbaric environments.

Funding for the acquisition of state-of-the-art equipment for biotechnology research must be made available. In addition, two related issues faced by academic institutions must be addressed:

- Universities have trouble obtaining funds for equipment maintenance.
- Universities have difficulty taking advantage of Federal instrumentation grants which require matching funds.

Repositories

New and expanded repositories are needed to hold and preserve the many specimens and large amounts of data crucial to biotechnology research. There are many acute needs:

- Germplasm for agricultural research must be collected, characterized, and maintained.
- Genetic stock centres are needed for marine organisms and other nontraditional organisms. The wide range of organisms apart from those traditionally used in laboratory research represents a biotechnology resource that must be developed if the United States is to maintain its strong leadership in biotechnology.
- Repositories must be developed for recombinant DNA reagents generated in genome projects as well as for specialized purposes.
- Repositories are needed for expanded microbial collections and particularly to store mixed cultures that prove promising as bioremediation agents. Current repository practices focus on the storage of pure cultures; the requirements for long-term storage of mixed cultures are poorly understood and must be developed.
- Repositories for marine organisms are needed to make these organisms accessible for biotechnology research. Genetically homogeneous strains of selected organisms at various developmental stages — embryos, larvae, juveniles, and adults — should be produced and made available on a year-round basis.
- International efforts should be made to coordinate culture storage and collection activities so that international standards for technical information can be developed and the exchange of data and cultures maximized.

Databases and Reference Standards

A critical issue in biotechnology is how to manage the massive amounts of information generated and ensure that user communities have access to the data in a useful form.

The solution lies in development of publicly accessible, relational databases, such as those that allow scientists to search across plant, animal, and microbial kingdoms for similarities in DNA sequences. These public databases open the door to discoveries about the fundamental nature of living organisms and ultimately to development of new biotechnology applications. The use of databases as integral tools in biology is in its early stages, and the existing database infrastructure is unlikely to provide an adequate resource base for the future. There are several specific needs:

- New and enhanced databases are needed to store and make available information about genomes, macromolecular structure, biological processes, collections of organisms, engineering design parameters, and other topics of interest to biotechnology researchers. These databases should be interactive, allowing manipulation of the data for process modelling and product design.
- International efforts in database development are essential, and information must be accessible to researchers from all nations. Interagency collaboration in support of computational sciences will help fulfil this need.

Reference standards for quality assurance of biotechnology-based measurements are also important. Reference standards assure accuracy and reliability in biotechnology research and commercialization and allow critical measurements to be compatible among different laboratories. Reliable production of high-quality data reduces the need for duplication of data.

Available reference standards for assigning measured values to proteins, glycoproteins, and DNA fragments are inadequate. Standards for measurement in quality assurance would have a major impact in a broad range of research areas, from genome projects to manufacturing. Quality assessment and comparability of results across different laboratories will be a rapid growth area and major tool for dissemination of information and technology transfer.

Chapter 2

Biotechnology in Agricultural Development

Introduction

We are at the threshold of 21st century, where great challenges on three fronts namely, production, pollution and population are awaiting the developing countries. Moreover, in the era of privatization, globalization and liberalization, there is heavy international competition within as well as between the countries on all fronts including the agriculture. The present scenario indicates that there would be "survival of the fittest". Needless to mention that by 2000 AD our population will be around one billion the priority is clear-cut before us. We have to climb upward to achieve higher and higher production to ensure that every mouth of our ever increasing population has enough to eat.

The quest for agriculturists for greater productivity and improvement of existing cultivars with better quality food continues with great vigour and spirit. The conventional methods of plant breeding and traditional agricultural practices have done a tremendous job and contributed to a great deal towards the above goal. However, in view of the acuteness of the problem and renewed fears regarding the availability of the proper and enough food, these methods alone are not sufficient to meet the situation.

Although, the FAO believes that food production will continue to increase for the next 20 to 25 years at the same rate as experienced during the past 30 years to meet the futuristic demand. The worldwatch Institute has, however, reported that this rate may not be sustainable.

Moreover, they believe that science and technology can no longer ensure the onward march of achieving higher and higher yields. In addition we are also subjected to an almost daily litany of doom and despair as a result of global warming, depletion of ozone layer, oil and water running out and so on. The next century certainly packed with trouble if the meteor does not get us first, then we will gradually either be roasted, frozen or desiccated, and to make doubly sure, shall certainly be starved as well.

It seems that if we have to stand any chance whatsoever in the next century, our capacity to adopt the living resources will be extremely important for our survival. Because of changes in the environment and depleting energy resources, this capacity to adapt will be every bit as important to those areas of the world which are blessed with overproduction like Europe today. Accelerated adaptation through genetic change is, of course, one of the important means by which plant breeders have achieved their aims in the past. Today the new techniques of biotechnology in general and genetic engineering in particular may offer to the plant breeders and agriculturists the chance of speeding up adoption to an extent hitherto considered impossible. This new power will undoubtedly bring risks but it could also be the only chance we have for escaping the dangers ahead. Biotechnology being a multidisciplinary subject requires coordinated efforts and expertise of scientists with different backgrounds to put together in order to achieve success. It may be defined as may technique that uses living organisms, or parts of organisms to make or modify products, to improve plants or animals or to develop microorganism for specific uses.

It includes well established technologies such as those used in breeding, plant propagation and conventional animal vaccine productions. Modern biotechnology encompasses the more recently developed techniques involving the use of recombinant DNA technology, monoclonal antibodies and new techniques of cell culture. Recent development in biotechnology have made it possible to move genes from microbes, plants and animal species into plants of interest. Biotechnology helps in the commercial utilization of gene and other things for the benefit of human kind. However, destruction of ecosystems has been a major concern of biotechnology.

Developing countries are naturally attracted to the potential applications of biotechnological research in solving problems of hunger, energy supply and improving the quality of life. The priorities of different countries however very widely, In this context, the National Biotechnology Board of India has chosen genetic engineering, photosynthesis, tissue culture, enzyme engineering, alcohol fermentation and immuno-technology as areas of immediate interest.

Application of Cell Biotechnology

Cell biotechnology or plant tissue culture is an enabling technology from which many novel tools have been developed to assist plant breeders. These tools can be used to increased the efficiency of the breeding process, to improve the accessibility of existing germplasm and to create new variation for crop improvement. They include micropropagation, another culture, in vitro selection, embryo rescue, somaclonal variation, somatic bybridization and transformation. Of these somaclonal variation occupies a somewhat unique position, because it has both the advantages and disadvantages of tissue culture systems.

Somaclonal variation was first defined as such by Larkin and Scowcroft who reviewed the subject in 1981 and were among several authors at that time to draw attention to its potential use for crop improvement (Larkin and Scowcroft, 1981). An extensive number of reports soon followed in a whole range of species, indicating that somaclonal variation was widespread, and therefore, ostensibly accessible to all plant breeders (Karp, 1991).

***Table** : List of varieties developed from somaclonal variation.*

Crop	*Country and Institution*	*New trait*
Geranium	Deptt. of Hort, Purdue Univ USA	Vigour and attractive flowers
Sweet potato quality of roots	North Carolina Res, Services, USA	Colour, shape and backing
Sugarcane	Sugar Research Centre, Fiji	Yield and disease resistance
Maize	Motecular Genetics, USA	Trytophan content
Tomato	DNA Pl. Tech. of New Jersey, USA	Dry matter content, Disease resistance
Rice	Plant Research Institute, Japan Unv of Agri. Sci., Godollo, Hungary	Yield Disease resistance
Celery	DNA Pl. Tech. of New Jersey, USA	Processing, Yield & efficiency
Brown mustard	ICAR, New Delhi	Yield.

The Utilization of new genetic variability has become one of the major objectives of tissue culture, The assembly of genetic variability is vital for improvement of crop plants. Somaclones for the resistance of downy mildew. Fiji disease and eye spot disease were identifies. Application of somaclonal variation in crop improvement can produce increased yield and resistance to biotic as well as abiotic a stresses. Some of the crops where somaclones were produced are rice, wheat, maize, tomato, geranium, sweet pototo, sugarcane, rice, celery and brown mustard. Our country research efforts have developed a variety called 'Pusa Jaikisan' in brown mustard evolved by somaclonal variation.

Micropropagation is another technique of great significance to agriculture which is based on in vitro culture of plant cell. Each and every cell of a plant has the potency to be regenerated into complete plant by tissue culture techniques and this property of the cell is known as totipotency. Using this technique hundreds and thousands of seedlings can be generated in a very short time, starting from a limited number of explants. This methods of multiplication has two major advantages. First is the multiplication of new elite cultivars. For example, after its release, a new sugarcane cultivar may take 5-10 years before sufficient planting material may be generated for wide coverage in the farmers field.

But using micropropagation techniques a comparable level of multicipation may be achieved within 2 years. The second advantage is that we get disease free planting material because of the use of aseptic conditions during tissue culture. There is also reports of rejuvenation and increased vigous following tissue culture. This can also help in rescue of important materials which are on the verge of extinction due to disease attack. For example, a rare variety of banana used for temple offering in Karnataka has been rescued through tissue culture techniques.

Micropropogation of fruit and ornamental plants provides a useful avenue for the employment generation for rural youths at cottage level industries. In species like Daffodies and gladioli, microporpagation techniques are being used to speed up the release of new varieties. In strawberry, millions of plants can be produced from a single mother plant in one year.

Micropropagation work undertaken at our Defence Agricultural Research Laboratory. Pithograh was resulted into successful multiplication of strawberry, petunia, carnation and tomato hybrid. In case of strawberry a well adapted variety of hill vi., Jyolikit Selection was taken for micropropagation studies with objective to large scale production of disease free propagules since this fruit plant is highly susceptible to number of fungal, bacterial and viral diseases. The result showed that internodal segments are most suitable explant material for micropropagation under in Vitro condition. MS formulated with 1ppm of BAP and 0.1ppm of GA3 was found optimal protocal for maximizing shoot proliferation in explant where MS media modified with 1 ppm 1AA was most suitable for early rooting in shootlets. Peat moss was found to be most suitable potting culture for hardening of tissue culture raised plants of strawberry var. Jyolikot Selection.

Micropropagation studies in Petunia Hybrida were undertaken in order to raise the plant seedlings in mass. The result showed that optimum number of shootlets could be obtained after culturing the leaf explant of size 5mm X 2.5mm in MS media supplemented with 1ppm of BAP + 0.5 ppm of IAA with an inclubation period of 34 days. Protocol standardisation revealed that MS medium supplemented with 1 ppm of IBA was suitable for an early and healthy rooting in tissue culture raised shootlets. Hardening process was most successful with high relative humidity (90-95%), Temperature 23 + 10 C and potting mixture of sand and soil in equal proportion.

Work on micropropagation of carnation was resulted into standardization of an vitro culture media for shoot and root proliferation. Protocol found most promising for shoot proliferation was MS medium supplemented with 1 ppm of BAp AND 0.1 ppm of IAA where MS medium supplemented with 1 ppm of NAA was identified to most suitable for rooting in shootlets.

Micropropagation studies were also undertaken in hybrid tomato to generate large scale seedings to fulfil the need of army personnel as well as local growers with pure and healthy planting material. Based on the observation it was found that tissue culture medium comprised of MS,0.5 ppm kinetin, 0.5 ppm IAA and 0.3 ppm GA3 was most suitable for shoot growth and MS medium supplemented with ppm of IAA or MS with 1 ppm of IBA was most suitable for early and healthy rooting in the explant material. Tissue culture regenerated plantlets showed high survival under relative humidity varied from 85-95 per cent, temperature 25+20C and potting mixture sand and oak forest humus in equal ratio.

Application of DNA Technology

Genetic engineering opens a totally new dimension for bioprospecting. The search for new genes and their application is the primary objective of the biotech industry. Gene technology now enable humans to integrate revolutionary new properties in to cultivated plants through inter-specific or inter-generic gene transfer which was not possible through classical approach of crop improvement.

Genetic engineering is the direct introduction into a plant of an isolated or modified single gene using transformation techniques. Now it is possible to transform almost all of the plants cultivated by man. Where this has not happened, it is more to do with the insignificance of the plant in agriculture or forestry rather than a reflection of a stubborn resistance to all attempts to transform it. A couple of years ago, cereals were regarded as being recalcitrant species. This has new changed and transformation has been reported as being successful for nearly all cereal species.

Transgenic Plants

Transgenic is used to describe plants which have had DNA introduced into them by means other than by the transfer of DNA from a sperm cell to an egg. Development achieved through modern biotechnology in the form of transgenics have given new options of controlling insects, diseases and weed pests, while improving the overall integrity and consequences of agricultural practices. Biotechnology combined with modern crop management techniques and the responsible use of pesticides gives the best tools available to ensure healthy, high yielding crops. However, the benefit expected from the release of genetically modified organisms into environment are quite more which are seems to pay major role in bioremediation, environmental improvement, agriculture, food industry and health care. A number of transgenic plants in various crop species covering wide range of altered characteristics have been approved for commercial cultivation.

Resistant to Biotic Stresses

Damage to crops by biotic factors like insects-pests and weeds is a major limiting factor in agriculture economic in tropical and temperate regions of the world. Despite large scale investment in the chemical control of pests, they cause great economic loss by damaging or destroying the crops. With a tremendous development in techniques to engineer crops genetically, we now have the ability to make broader use of natural insecticides. The choice of a Bt endotoxin, as the first insecticidal protein for introduction into plants, was based on the extensive knowledge gathered about this class of crystal protein since 1902 and such a strategy has been successfully achieved. The application of glyphosate kills crop plants just as effectively as it kills weeds. The usefulness of glyphosate as a weed control agent in agriculture could be enhanced if resistance to herbicide could be selectively engineered into crop plants. Transgenics have been developed by introducing the corresponding resistance genes from weeds and microbes into crop plants such as tobacco. In addition genes conferring resistance against herbicides, viz., Sulphobylurea, Imidazolinones, Phosphinothricin, Asulam, Bromoxynil and 2,4-D have been engineered and transgenic plants developed.

Abiotic Resistance

A large area of our country is under stress conditions i.e. saling, alkaline water, cold and heat stress etc. transgenics produced from introducing mtl 1D gene from E. coli in tobacco and arabidopsis showed tolerance to high leave of NaCl. Thermoc tolerance was found in Arabidopsis plants engineered with hsf gene. It has been reported that over-expression of sac B gene from Bacillus subtillis leads to high level of fructans in tobacco cells, and this is associated with increased drought tolerance whereas chemically synthesized antifreeze protein genelala 3 leads to improved freezing tolerance of tobacco plants.

Genetic engineering for improved tolerance to abiotic stresses is, therefore, the need of the hour as the existing cultivars in most cases are capable of giving much higher biomass that what we harvest today. While it is true that the response of plants to these stresses is multigenic, the recent success achieved in genetic engineering against excess salt, high and low temperature as well as less or excess water by altering individual gene is noteworthy.

Product Quality

A group of scientists have reported the ethylene synthesis in transgenic tomato plants, using anti-sense to a gene that has been identified as ACC oxidase. The enzyme convert ACC to ethylene. Ethylene production was inhibited by 97% with a signification reduction in over-ripening and shrivelling. Flavr Savr and Endless Summer in tomato, Freedom II in squash, high lauric acid reposed (Canola) and Round Up Ready soybean are some examples of the crops that already being commercially grown in developed countries.

Genetic engineering of metabolic pathways of fruits and vegetables has therefore given control over post-harvest process which not only leads to improvement of quality characteristics but also enhances processability, transportability and prologation of shelf like.

In addition, engineered plants with change fatty acid composition of edible oil, reduced antinutritional components in many food crops, production of palm oil or animal derived industrial oil into plants like Brassica are some of the good examples of novel products that can be expected from biotechnology. Significant progress has been achieved for production of biodegradable plastics derived from biomsass of genetically altered plants having gene from bacteria. With the advancement in DNA technology and understanding the structure and function of gene and its products more and more transgenics are coming out from various research programme all over the world. However, boon of biotechnology has been confined to few developed countries and developing countries are still in vogue to harvest the benefits of modern technology. More that 48 transgenic has picked up remarkably in last couple of years. During 1986-87, 25000 field trials were conducted with transgenics of more that 60 crops in 45 countries. Sixty per cent of these trials were in first 10 years and 40% in the last two years. Area under transgenics cultivation has also increased tremendously. During 1996, the transgenic crops covered an area of 2.8 million has which increased about 4 times during 1997 and 10 times during 1998.

Table: *Area of transgenic crops planted (m ha)*

Country	***1997***	***1998***
USA	8.1	20.5
Argentina	1.4	4.3
Canada	1.3	2.8
Australia	0.1	0.1
Mexico	<0.1	0.1
Spain	0.0	<0.1
France	0.0	<0.1
South Africa	0.0	<0.1
Total	11.0	27.8

Source: Anne Simon Maffat (1998)

Conclusion

A major challenge for agriculture in the 21st century in developing nations would be to speed up production process economically to fulfil the aspirations of huge populace, to achieve diversification and adding value to the primary produce so as to make agriculture enterprise farmers as well as environmental friendly. Biotechnology based advance technologies are expected to materialize many of our expectations in the coming millennium.

Viruses

To minimize the losses due to viral disease and to reduce the use of chemicals, the cross protection phenomenon of viral coat protein and movement proteins have been employed. This method has been used to develop transgenic plants which conferred significant levels of resistance in tobacco plants to TMV or to alfa mosaic virus (AMV) using the AMV coat protein gene.

In some viruses it has been found that certain sequences termed satellite RNAs act to reduce disease symptoms in infected plants and as such it is being utilized for developing transgenic with virus resistance. Antisense RNA which binds to sense RNA and artificial ribozyme molecules which cleave specific gene transcripts were found very powerful tools for the inhibition of viral disease development. Genetic engineering of antisense or ribozyme genes complementary to viral genes whose products are essential may prove one of the best way of viral disease control.

Insects

Transgenic crop plants that produce pesticidal proteins such as Bt toxin can offer the growers advantage of increased yields, numerous environmental benefits due to decreased use of conventional insecticidals. A transformed tobacco plants with the gene encoding the cowpea trypsin inhibito, in feeding experiments demonstrated that expression of the gene conferred resistance to a range of insect genera, including Heliothis, Spodoptera, Diabrotica and Tribolium the protein is not however toxic to humans.

Herbicides

The use of herbicidea has become an important feature of modern agriculture as a means of controlling weed species, however, it is economically feasible to produce specific herbicides for use with every individual crop species. A more generic approach would be to transform a range of crop plants with a gene which confers tolerance to a single non-selective herbicide.

Biotechnology for the Cotton Farmer Problems and Prospects

Cotton or white gold as it is aptly called is grown for its lint and seed which yield cotton fibre and seed oil, respectively. This crop occupies 32-33 mha of world area with a production of 20-25 metric tonnes. In India its area spans over 7.5 mha with an average yield of 290 kg/ha of lint and 870 kg/ha of seed cottom. To meet the challenges of 2000AD with a population of more that 960 million, a total production of 19 million bales is required as against the 13-14 million bales of today. This can be achieved by the use of improved crop production practices coupled with appropriate pest management tactics.

In addition, generation of novel. Transgenics may help achieve the near impossible. Genes that have been identified as potentially profitable, if engineered into acceptable cultivars can be used to generate such transgenics. Among these are genes imparting resistance to herbicides, insects, pathogens and abiotic stresses. It is also widely accepted now that a number of other qualitative characters can be improved, such as fibre strength, fineness, colour and thermal adaptability of the fibre.

Thansgenic plants have become realistic components of stress management world over. Bollworm and herbicide resistant transgenic cotton have received the approval of the Environmental Protection Agency and have been commercially released in the US for cultivation. Considering the fact that numerous biotic and abiotic stresses limit cotton production, it is likely that future strategies might orient towards the development of a multiadversity resistant high yielding transgenic cotton variety with superior fibre qualities.

The most important aspects in the development of transgenic plants are

- Identification, isolation and clining of the desirable genes
- Transformation of plant cells or tissues with a suitable vector
- Regeneration of the transformed cells or tissue
- Confirmation of the gene integration and expression
- Hardening and field establishment of the transgenics.

Major achievements worldover in the field of cotton transgenics

Genes

Genes for Insect Resistance Used

- Toxin genes from Bacillus thuringiensis
- Protease inhibitor genes from plants & the horn worm

Genes used for herbicide resistance :

- Nitrilase gene from a bacteria Klebsiella spp. for Bromoxynil resistance

- Mono-oxygenase gene from a bacteria Alcaligenes eutrophus for 2,4-D resistance
- Acstolactate synthase gene for resistance against sufonyl urea & imidazoline group
- ERPSP synthase gene for glyphosate resistance.

Genes used to Improve fibre qualities :

- Polyhydroxybutyrate & polyhydroxyacetate from bacteria
- Expression of indigo pigment in fibres

Genes to be Used

- Cholesterol oxidase genes against insects
- Heliothis stunt virus genes against Heliothis
- Genes encoding chitinases glucanases, attacins cecropins vital coat proteins etc. against diseases.

Transformation

- Agrobacterium medicated transformation
- Particle acceleration gene delivery systems have been successfully used all over the world.

Regeneration

- Callus fissue regeneration-mostly genotype specific
- Protoplasts were regenerated successfully.

Meristematic tissues were successfully regenerated.

Expression

- Enhanced expression was achieved using truncated forms of the full length genes, repeat copies of the propoter, modification of the bacterial codons to plant prefered type.

Commercial Release :

- Bollworm resistant transgenics-NUCOTN 33M & NUCOTN 35B were released in the US and cotton transgenics-are in now under field cultivation in Australia, South Africa and China.
- 8 lakh hectares were planted with the transgenic cottons in 1996 and 1997
- More than 8,000 hectares were damaged by bollworms.
- Transgenic cotton plants in Australia were also found to be attacked by the bollworms.
- Herbicide (Bromoxynil) resistant transgenics BXN 57 and BXN 53 were released in the US in 1996.

Problems Encountered/Expected

- Gene silencing and expression instability
- Resistance in insects to the toxins used. Studies on baseline toxicity indicated that there is a natural variation in the ability of Helicoverpa armigera in tolerating the CrylAc toxins. Some strains are capable of surviving the totoxins by virtue of an in built capacity to tolerate the toxins, while others are not. While development of Helicoverpa resistance to the CrylAc toxins itself may not be of immediate concern, a slight shift in the tolerance levels resulting from continuous exposure to the transgenics, can cause control difficultes.
- High cost of transgenic seed.
- Microecological changes resulting in pest shifts. CrylAc is a broad spectrum lepidopteran toxins that can cause significant changes in almost all populations of lepidopteran insects occurring on the cotton crop. Some of these insects, which cause negligible economic damage to the crop, harbour parasitoids which have the potential to keep the bollworm populations under check. Consequently, in the absence or low populaions of natural enemies, the bollworms can emerge as stronger pests that before. Insects pests such as Spodoptera litura which are less affected by the CrylAc toxins and which have been under check due to the use of pyrethroids can resurgace as major pests as pyrethroid use is likely to be reduced on transgenic Bt cottons.

Some Recent Development in Cotton Transgenics

- Three Bt transgenic lines, 95-1, RIOI and R-19 have been indigenously developed in China through Agrobacterium medicated (Somatic embryo) and pollen tube pathway transformation. Thestability of Bt gene expression and field efficacy have been confirmed.
- Microprojectile bombardment parameters for pollen mediated transformation and genetype independent protocols have been develop in China.
- A new methods of transformation through injection of Agrobacterium into developing embryos, has been reported from South Africa.
- Stable transformation and regermation has been reported for Bt transgenics in Uzbekistan, China, Egypt and Australia. Pakistan reports development of cotton transgenic plants resistant torearcurl virus.

- An antisense DNA of CLCUV DNA-A bome ACI gone along with the antesense DNA of the AC2 and AC3 gene was used for the vector construction and transgenic cotton resistant to the CLCUV was developed in Pakistan.
- In Russia, A glucanase gene has been isolated from thermo philic bacteria and is being used as a new reporter system fortransgenic cotton development.
- Genetically engineered Baculovirus ACMNPV with genes from straw itch mite has been developed been developed by Monsanto and Zeneca with researchers from Madison, USA.
- Bt rransgenic cotton currently occupies 20% (85,000 ha) of cotton area in Australia. Isecticide use has been reduced by about 60% The efficacy has been moderate. The transgenic plants were found to express Bt proteimns only till the 95th day after which fluctuaions were observed. The reduction in expression was primarily due to down regulation which post transcriptional changes of the unstable RNA It was also reported that cotton tannings which increase with growth phase, act as antagonistswith Bt toxins.
- After two years of field cultivation, true resistance to Bt has not yet been detected in Australia.
- Reslistanor to Bt in field populations of Helicoverpa armigera is reported form China
- Annually 100 million dollars are now being spent in the US, only in search of new insecticidal genes.
- In the US, 45% of cotton area this year will be under transgenic cotton (herbicide and insect resistant). However, the area under Bt cotton seems to on the decline.
- Transgenic cotton developed in Australia using the Stunt virus genes, is not performing well.
- Transgenic cotton developed in the US using lectin genes, is not performing well against Heliothis virscens.
- The Australian report that Helicoverpa armigera is at least 100 times less susceptible to Cry toxins as cinoared to Heliothis virescens.
- The Window strategy for insecticide resistance management in Australia is still important for transgenic cotton.
- Cotton Pest spectrum in the US is getting altered after the introduction of Bollgard. Unsprayed Bt cotton sustained 4 times

more attack of tarnished bugs, 2.4 times more with boll weevil, 2.8 times more with stink bugs and Spodoptera.

- Due to these changes in pest complex farmers had to spray 3-5 times on boligard as compared to 6-8 times on non-Bt cottons.

Prospects and Potential

The application of biotechnology in cotton farming can be either in the form of production of fermentation products or novel recombinant products for use (example, recombinant Pseudomonas expressing CrylAc) or as transgenic plant with in buitt resistance to biotic and abiotic stresses. Transgenic crops with in built resistance to insect pests and diseases can be extremely useful as this would result in the reduction of insecticide use apart from making pest management simple for the farmer. Introduction of the bollworm resistant transgenic cotton is expected to reduce the use of chemicals used to control bollworm especially Helicoverpa importantly at a time when resistance to most insecticides including pyrethrolids is on the increase all over the world. Even if the introduction of the Bt cotton in India could result in a 25-30% reduction in insecticide use on cotton, this would mean a benefit of about of about Rs 300 crores, apart from the f-avourable impact on the environment so far, transgenic plants have been produced in about 60 plant species. Cotton has received special attention of the biotechnological companies in the developed countries who were attracted by the profit motives associated with the high value added to the transgenic seeds.

The firs Bt transgenic cotton has already been released, as Bollgard. The Delta and Pineland using DP 5410 and DP 5690 as recurrent parents. The D and PL brand BT transgenic were labelled as Nucotn 33B Two million acres were planted in 1996 in mid-south region of USA, mainly in regions where Heliothis virescens was problematic. Nearly 1.2 M acres were under Nucotn 33 B and the rest under Nucotn 35B.

The Bollgard brand Bt cotton seed was sold at US $34-36 as compared to US $8-9 per hectare for non-engineered cotton seed. The average cost of control of cotton insect pests in the US was approximately US $150 in the early 1990s. Hence the prices were still found to be attractive. The transgenics were found to be more effective against Helothis virescens as compared to pectinophora gossypiella and Heliothis zea.

Farmers growing transgenics had to sign'an agreement with D & PL stating that they would not keep seed for planting next year. Quick

ELISA tests have been devised to test for Bt toxins in plant parts, to check the illegal spread of transgenics.

Two more transgenic cotton varieties tolerant to herbicide Bromoxynil, trademarked as BXN 57 and BXN 53 were developed from the recurrent paren Coker 315 BYTHE Stonevuille pedigree company in collaboration with Calgene Inc. This seed was priced nearly 1.7 times the straight varieties.

In Australia the CSIRO and Narrabri research centre, in collaboration with Monsanto have developed transgenic cotton plarts that were commercial released in 1997. This introduction resulted in 50-60 % reduction in the US $ 93 million spent by farmers each year on insecticides. Pakistan which uses 90% of its total insecticides on cotton alone and China which is one of the major consumers of insecticides on cotton are reportedly strongly considering the prospects of introduction of Bt transgenic cottons.

Cotton acreage in India is only 5% of the total cropped area, yet it consumes more than 50% of the total insecticides used in the country. The pattern of usage, however, is not uniformly, spread, for instance in Andhra Pradesh alone, where cotton cultivation is only 0.3% of the total cropped area of the country, accounts for 17% of the pesticide use on cotton in India. Again within the average picture of the state, coastal AP uses 30% more pesticide than the state average. The approximate estimates of insecticide sales in Guntur district alone, where cotton is grown in 0.15 mHa, is about US $ M. Clearly areas with maximum pecticides use per hectare, necessitated mostly due to bollworms, are likely targeted market niches for the bollworm resistant transgenic cottons. Moreover, since the price of the transgneic seed in the US is nearly four times the high input cotton cultivators of the irrigated belt Growing Bt transgenics and herbicide tolerant transgenics on marginal land is not recommended by the company as these would not show any impact in the absence of strong insect pressure and broad leaved weeds. In most areas of rainfed cottons the use of insecticides and pest pressures are low hence, it may not be economical in such belts, to encourage the use of transgenics.

Now transgenic production technology for many crops has become an established routine procedure in many countries including India. The research on transgenic cotton in government funded research labs in India is almost in the final stages. Meawhuile Mahyco-Monsanto biotech is ready to market Bt transgenic cotton (bollgard) in India. The introduction of transgenics through these companies are to be in the form of FL seeds wherein Bt genes are inherited from the transgenic exotic germplasm. Such seeds may also be priced at a premium and it remains to be seen how enthusiastic the response of Indian farmers would be to the expensive input. Transgenic releases from government organisations would definitely enforce a competitive pricing to restrict the monopoly of the private companies.

The transgenic technology is also likely to be extremely beneficial to the private sector. For instance, a transgenic with resistance to a specific herbicide marginalises the use of other herbicides for chemical weed control and the chemical company holding the patent on the herbicide might be prove to be beneficial to the resource rich farmers. Clearly, here the targeted market is primarily high input farmers. The monopoly of transgenics due to patenting rights held by the biotech companies may also lead to high pricing of the seeds.

A Significant socioeconomic issue that arises from the introduction of transgenics into the Indian farming system is that the high priced seeds may benefits the prosperous and large farmers thus providing a negative externality on small and marginal farmers. On the other had it is also argued that the developments from the application of biotechnology would be beneficial to low input farming practises wherein the cost of chemical inputs can be minimised. It is now only a matter of time before we experience the full social. economic and environmental impact of transgenics in our country.

Biotechnological Tools for Increasing Productivity of Pulse Crops

Pluses play a significant role in Indian agriculture because they provide protein rich diet to poor people of the country. They contain 20-29% protein i.e. about 2.5 times more than cereals. Inspite of release of large number of high yielding varieties, extension of irrigated areas and use of agrochemical, growth recorded in total production of pulses from 11.82 million tonnes in 1970-71 to 13.19 million tonnes in 1995-96 is considered to be non revolutionary. This resulted in decline in per capita availability of pulses to 37g/person/day. There are several reasons for stagnation in pulses production such as narrow genetic base, lack of genetic variability in useable genepool, inability to transfer increased input to increase output (seed yield) susceptibility to large number of biotic and biotic stresses, photoperiod sensitive and non synchronous flowering and maturity. It has been estimated that on account of insect pest complex infesting pulse crops alone, nearly 2.0 to 2.5 million tonnes of pulses is lost annually. The monitory value of this loss will be approximately Rs.3000 to 3750 crores.

Conventional methods of crop improvement have not improve productivity of pulses with significant pace due to several reasons. These methods are slow, expensive and time consuming. Besides lack of reliable resistant sources, undersirable linkages and long gestation period are considered to be the main limitation of these methods. Hence, there is dire need to introduce innovative biotechnological approaches to enhance pulse productivity to meet out the market demand.

Biotechnology and genetic engineering has emerged as potential tool in recent year for genetic manipulation of crop plants. The said technology promises to improve crop productivity through complementing traditional breeding by decreasing dependence on harmful chemicals pesticides. Fertilizers and antibiotics etc. The tools of biotechnology are now available to plant breeders and can be used effectively to supplement conventional practices in improvement of pulse crops. It is expected that products of biotechnology will be ecologically sage, economically feasible and technologically expectable to Indian farmers.

Effort have been successful to develop several intergeneric and interspecific hybrids using embryo/ovule/ovary/endosperm culture in grain legumes, Several useful genes imparting resistance to various biotic and biotic stresses have been successfully introgressed on wild relatives of lathyrus, lentil, chickpea, pigeonpea urd bean and mung bean. The resistance to Aschochyta blight to chickpea has been successfully transformed from wild species Cicer judaicum to elite chickpea genotype PDG 84-10 at IIPR, Kanpur.

The techniques of in vitro mutagenesis in vitro selection and somaclonal variation are being also successfully employed to various pulse crops to develop promising genotypes resistant to various biotic and a biotic stresses. This technique has resulted success in developing Aschochyta blight resistance in chickpea. Further early results in vitro selection for salinity tolerance in chickpea is also very encouraging at IIPR, Kanpur. Similarly, the technology of cell suspension culture followed by selection of mutants has also been used successfully in major pulse crops at different centres in the country.

Identification and characterization of pathogens and pests is problematic and always pose challenge to plant pathologist and breeders in varietal development of pulse crops. The techniques of molecular diagnosis and DNA finger printing has facilitated resulted easy diagnosis and the precise characterization of pathogeny isolates and insect biotypes at molecular level. This has increased the efficiency of breeding programmes. Important pathogent like Aschochyta rabi and

Fusarium oxisporum of chickpea have already been characterized using DNA fingerprinting techniques.

In a new world order and changed senario the characterization of germplasm and plant material is very essential to protect out natural wealth. The techniques of DNA fingreprinting have been proved very handy in protecting the intellectual property rights. These techniques are now being successfully used in pulse crops to characterize and catalogue the accessions of gene bank to protect the genetic wealth.

DNA marker techniques are the new innovations and have tremendous potential in increasing the efficiency of conventional breeding programmes. DNA marker technique can be used for tagging important genes which can subsequently be used for early, easy and precise selection in segregating of promising plants populations. DNA markers have been found linked to wilt resistance in chickpea, powdery mildew resistance in peas and nemarode resistance in mung bean and urd bean.

Transgenic technology which involves insertion of foreign DNA sequence has tremendous potential for improvement of crop plants. Efforts are on to develop transgenic plants resistance to pod borer in chickpea and pigeroxpea using Bt. Crystal protein gene from a soil bacterium. Early results obtained at various centres in the country are very encouraging. Similarly, efforts are also being made to develop transgenic urdbean and mung bean using coat protein gene. Current status and future strategies for genetic manipulation of pulse crops using biotechnological tools will be discussed in detail.

Biotechnological Applications for The Improvement of Indian Tobacco

India is one of the top 10 countries cultivation tobacco for the part ten decades both for domestic and international market, India is the third largest producer of tobacco in the world with an output of approximately 520 m kg. Of which FCV tobacco account for 120 m kg. The economic importance of to Indian exchequer is peogressively increasing in the last 10 years. Tobacco generates 10% of total excise revenue and tobacco 4% of the value of total agricultural exports, India has the unique distinction of growing different types of tobacco in different Agroclimatic zones.

Tobacco provides direct and indirect lively-hood to 26 million in India including 6 million farmers and workers. Many small and marginal farmers were benefited by growing tobacco. Research efforts to identify alternative crops to tobacco indicated that no other crops is as remunerative as tobacco. Stability of the worldwide demand of the crop, the crops hardiness and ability to grow in climate and soils non-suitable for other crops, relative ease in transporting tobacco and tobacco's profitability compated to other crops are incentives for farmers growing tobacco thus making tobacco one of the most appealing cash crops to farmers.

Tobacco crops is considered as major employer worldwide. Despite technological progress tobacco cultivation remained as one of the most labour intensive crops among all other agricultural crops. Women in particular and skilled labourers, who would otherwise have little chance of employment are able to earn living by working in tobacco cultivation.

WHO in its report has acknowledged that in developing countries tobacco can create vast employment. In view of its great role in the economy of the country and farmers, further improvement of tobacco crop through Biotechnological tools will benefits them a great deal.

Biotechnological applications: Scientific system of tobacco cultivation in India was introduced with the establishment of Central Tobacco Research Institute (CTRI) at Rajahmundry in 1947. Tobacco research in India over the past 50 years through CTRI has made significant contributions to the economic upliftment of tobacco farmers by identifying high yielding varieties and suitable package of practices This is reflected in the good reputation for Indian Tobacco in the International market, as leaf of desired quality traits as per the specific requirement of traders is produced.

Initially, conventional breeding methods were utilised for improving the yield and quality of tobacco. However, later application of biotechnology tools for supplementing the conventional tobacco improvement programmes or initiated.

Different biotech techniques like anther culture, fertilized ovule culture micropropagation, protoplast culture somatic hybridization and also genetic engineering are being utilised. These techniques are helpgul is strengthening the conventional breeding programmes where ever they are week and failing to serve the purpose. Progress of biotechnology world, in different aspects, at CTRI is given below:

Yield Improvement

Yield improvement through conventional breeding programmes takes fairly long time. This can be reduced substantially (3-4 years) by culturing anthers (Anther culture). Of elite lines from early breeding populations in tissue culture. Through anther culture several dihaploids are produced and evaluated every year. An improved line D1 was developed, in this way, which is superior in yield. Efforts to incorporate budworm resistance through someclonal variation in this line is in progress. Stable resistant D1 someclones are there in field testing stage.

Resostance to Biotic and Abotic Stresses

Wild species of tobacco possesses resistance to various tobacco pests and disease. It is difficult and in some cases impossible to transfer this resistance from wild species to cultivated tobacco through conventional methods. In such cases, inter-specific hybrids were rescued through fertilized ovule culture and mass multiplied for further testing by micropropagation, In this way, interspecific hybrids possessing resistance to leaf spot diseases, black shank, root knot nematodes leaf eating caterpillar, aphids, budworm etc. were developed and evaluated for their resistance. Thus, stable pest and disease resistant lines were identified through Biotechnological tools. Efforts to transfer desirable characters from wild species through somatic hybridization is in progress.

Somaclonal variation was attempted to fortigy tobacco lines resistance to white fly (Bemisia tabci Genn.) (leaf curl) and budworm (Heliothis armigera Hb) and the results are encouraging. Somaclonal variation is highly effective to create variation when desired variation is impossible to get through conventional breeding methods and also variation is required only for one character.

Transgenic Bt tobacco cultivars processing resistance to leaf eating caterpillar (Spodoptera litura F) and budworm (Heliothis armigera Hb) were developed. Production of transgenics will give scope for transferring desirable characters from different species and genera.

Preliminary attempts were made to screen lines for drought tolerance in tissue culture. Such efforts will save expenditure on extensive field evaluation and work can be carried out in the off season also. Transgenic tobacco having wheat desiccation tolerant protein were evaluated water stress tolerance.

Herbicide Resistance

Orabache cernua os a root parasite of tobacco causing heavy loss in yield and quality. Glyphosate at lower does was found to kill Orabanche, but it is phytotoxic to tobacco. Hence research on production of herbicide (glyphosate) resistant tobacco lines through tissue culture was initiated. When farmer grows herbicide resistant lines in the field he can use berbicides without any harmful effect on tobacco, for the effective control of weeds and Orabanche.

Conservation of germplasm: Germplasm is the valuable source for various characters that are useful in future breeding programmes. This can also be conserved under in vitro conditions for long time. Also lines that cannot be maintained under field conditions due to non-flowering and sterility can be effectively maintained under tissue culture conditions. Interspecific F1s, mammoth mutants, in vitro rescued hybrids protoclones, BT transgenics various mutants are routinely being maintained.

Future Thrusts

Biotechnology can help Indian tobacco farmer in many ways, Particularly in tobacco improvement research and seed purity disputes, if offers a large scope. Research on tobacco improvement benefit farmer as he can get high yielding and pest and disease resistant tobacco for cultivation. In research various aspects like in vitro conservation and molecular characterisation of germplans, isolation, characterisation and transfer of desirable genes from across genera, molecular aided selection in breeding programmes, someclonal variation can play a great role. Utilising Biotechnological techques tobacco lines that can yield higher amounts of phytochemicals which are useful in the production various drugs and pesticides can also be developed.

In recent times many disputes about seed purity among farmers are coming up. At present morphological criteria is only being used for varietal identification. As morphology is highly subjective in nature and also varies with the type of environment the plant is grown, it is becoming highly difficult to solve the disputes. Under this situation molecular characterisation of cultivated varieties will help to identify correct varieties beyond doubt.

In future, giving emphasis on the above said lines will go a long way in benefitting the India tobacco crop as well as the farmer. However, as biotechnology got wider applications, any thing that put the farmer and environment at disadvantage need to be discounrages.

Biotechnological Approaches to Weed Management

Weed management is essential for good quality and quantity of food production. The presence of weeds in general may reduce crop yield by about 30% or more. Being potential pests, weeds have to be removed employing mechanical, chemical cultural and biological means. Weed management through application of chemicals (herbiciedes) is, through effective, not eco-friendly. The herbicides may pollute the environment and many alter the natural equillibrium. They also endanger and alter plant and microbial biodiversity. Effective non-chemical weed management may include various biotechnological approaches using plants and/or their products. These are useful both for the natural equilibrium and the environment. In addition they also condition the soil and improve crop production. Some of the recently evolved biocontrol and biotechnological methods of weed management developed and tried at the National Research Centre for Weed Science, Indian Council of Agricultural Research, Jabalpur are outlined in the following text for ready use for trials and for working our their cost effectiveness.

Management of Phalaris minor in wheat by using Fungi

At present Phalaris minor is one of the most predominant weeds in wheat fields. The Weed spreads through seeds. The weed has developed resistance against popular herbicides like isoproturon. Biotechnological approach using fungi offers an alternative possibility of management of the weed through biocontrol.

(i) Control of Phararis minor using Trichoderma viride: Trichoderma viride is used as a seed dressing fungi in various crops including in wheat for management of seed borne/soil borne pathogens. Interestingly, the fungus has been found to inhibit germination of seeds of Phalaris minor. Application of Trichoderma, viride grown in saw dust and neem oil cake as a base consistently inhibits germination of phalaris minor seeds and reduces vigour of the seedings of the weed. At the same time it improves yield of wheat by about 5%. The fungus has potential of controlling phalaris minor. Soil application of the fungus, wheat seed treatment and spray of the fungus were all effective in controlling the weed.

(ii) Control of Phalaris minor by gilocladium virens: Control of Phalaris minor by ?Gliocladium virens has also been found to have potential for control of Phalaris monor by inhibiting seed germination in a similar manner as does the Trichoderma viride.

Management of Parthenium (Parthenium HysterophorusL) by Marigold (Tagetes PatulaL)

Parthenium is an obnoxious weed of worldwide occurrence, harmful to human and animal health, agriculture, environment and the natural biodiversity. The weed reduces yields of crops and grasses. When shown in parthenium infested field or area, marigold inhibits seed germination and growth of the weed. It also reduces population of the weed. In the next generation, reduction of parthenium population is as height as 85-100%. This has been found primarily due to competition and through all elopathy (affecting parthenium plants by releasing chemical substances including thicophene into soil through roots.) Built up of the phytotoxic substances of marigold origin prevent parthenium seed germination, growth of plants, and flower and seed production completely in 2-3 years. This technology appears to be economically rewarding for marisgold flowers have good market. This can be achieved simply by spraying marigold seeds over parthenium infested area. The technology is simple, effective, economically rewarding, and ensures self perpetuating control of parthenium in wastelands. Marigold to parthenium plant ratio of 0.5 to 4 appear effective for near complete control of parthenium. Adequate moisture availability in soil facilitates establishment of marigold in parthenium infected area.

Management of Parthenium by fungi: Two fungal species namely Gliocladium virens and Trichoderma viride which have been widely used as seed dressing to present parthogenic damage have also been found effective for parthenium control.

(i) Control of parthenium by Gliocladium virus: Glioladium vieus and 10% neem oil separately as well as in combinations spray control pathenium and other broad leaf weeds like chenopondium album, Melilotus alba and Medicago sp. In wheat crop. The fungus could be explored as a self perp control measure for non-cropped area also.

(ii) Control of parthenium by Trichoderma viride and need oil: Thrichoderma viride and neem oil separately as well as in combinations spray control parthenium under field conditions. The fungi has potential of becoming a self perpetuating means for control of parthenium under non-cropped area. Trichoderma viride also acts as an antidote for parthogens of parthenium that can possibly be used as biocontrol agents. If any of the parthogens attack crops, the Trichoderma virid can be used to overcome the problem.

Management of Parthenium by Utilising as a Source of Nutrients

Parthenium is able to extract nutrients even from nutrient deficient soils in which it grows. It has very height levels of nitrogen (3%), phosphorus (0.2%), potassium (4.5.%) and other macro and micro-nutrients. Uprooted, dried and cut to small pieces or powdered parthenium plants at preflowering stage can be applied directly to crop fields. Apart from increasing crop growth and yields, the parthenium residue conditions soil with the organic supplement. Keeping safety considerations, it may serve as an incentive for participation of public in its ecofriengly control.

Weed management through mid-summer irrigation and covering the beds with black polythene: Weed problem in most cases is due to seeds of weeds present in soil seed bank which germinate before, with or after planting the frops. Mid summer irrigation and covering the beds or fields with black polythene thoughout the summer kill most of the readily geminable seeds in soil seed bank. Polythene covering can be removed after summer before planting the crops.

The treatment may provide near complete weed control. This technique is suitable for areas where summer temperature exceeds 400 c. Weed management through black polythene stripe mulching Where crop is grown in rows, inter row space could be covered with black polythene mulching. This kills weeds appearing in the area due to interception of light necessary for survival of the plants. The are many other ways of biotechnological management of weeds such as bioling water jet treatment of soil, by intensive cropping, by increasing plant density and chosing allelopathic crop species etc.

Biotechnologycal Application

Techniques such as embryo transfer, dilution of semen and its cryo preservation and artificial insemination (A.I.) are well developed and expertise in these fields is available. This technology is extensively used in the case of cows and to some extent Buffaloes. The main lacuna in the case of sheep and goats is the non availability of improved breeds which could be utilised. As these animals are owned by the poor strata of society and there is a mistaken prejudice that they are responsible for environmental degradation no sustained efforts have been made to improve indigenous breeds and thus make available genetic stock for upgrading. In spite of this the numbers of these animals have been increasing rapidly (*2.5% per year in the case of goats) due to an increasing demand for meat. Our breeds have developed over the centuries to survive and adapts themselves to harsh environmental conditions but do not universally possess ability to respond to better management and nutrition.

Outstanding work in the case of the Sirohi goat was carried out from 1981 to 1993 by the Indo-Swiss Goat Development and Fodder Production Project in Rajasthan where they showed that outstanding animals are available within this breed. However, before the impact of this project could be felt the project was discontinued.

Goats

Out institutes is a very small NGO with only three technically qualified staff, one of who is in an honorary capacity and one on deputation from the Government of Maharashtra. We decided to initially concentrate on crossbreeding using the outstanding South African Boer Goat the worlds only meat goat as an improve breed. With assistance from the Government of Australia we imported 20 embryos of this breed from Australia implanted them in local goats and now have a small flock of Boer goats.

This goat unlike many temperate breeds of Dairy goats have proved to be extremely hardy, producing excellent crossbreed when crossed with local goats. In order to utilise this germplasm effectively we have developed and modified the techniques of semen cryopreservation so that it can be done with low cost equipment producing pellets instead of straws. We have also improvised. A.I,. techniques using cervical insemination to get better conception rates. We have so far trained over 50 trainees among them farmers. We have supplied pure Boer goat frozen semen to Governmental agencies in Orissa, Tamil Nadu as well as to may private goat rearers in Maharashtra.

Our veterinarian is capable of carrying out multiple ovulation and embyo transfer if there is need for rapid expansion of our flock. However, this technology is expensive and only to be used if costs justify its use.

A I technology in goats has been spread to near by villages where fresh diluted semen as well as frozen semen have been used to produce 800 crossbreeds over the past five years. This technology is viable and easily transferable but unfortunately dependent on the sincerity of the inseminator.

Sheep

We were unable to identify a suitable improve breed to be imported which had the most desirable but greatly neglected attribute of prolificacy. We did locate a prolific and unusual breed in the hot humid sundarbands of West Bangal called the Garole. Up to 1993 this breed had remained unnoticed by the community of sheep breeders. We did however import a small flock of Awassi Dairy Sheep from Israel. This breed has produced an outstanding crossbred animal when crossed with the Malpura of Rajasthan. This crossbred has been developed by the Central Sheep and Wool Research Institute (CSWRI) using Awassi rams obtained from us. The Australian Centre for International Agricultural Research has sponsored a collaborative project between the University of New England, Australia, The national Chemical Laboratory, Pune and the Nimbkar Agricultural Research Institute (A SISTER NGO of Maharashtra Goat and sheep Research & Development Institute), Phaltan entitled," Prolific Worm Resistant Meat Sheep for Maharashtra, India.

This project is making full use of Biotechnology. We are comparing three breeds of Indian sheep the Grole, Deccani and the bannur. We hope to evaluate lamb production including prolificacy and resistance to parasites and try and determine the genetic basis of these differences by DNA analysis. We also hope to account for the interaction between measurement of resistance to parasitic nematodes and environmental effects. The project has completed one year and the results so far are encouraging.

Our ultimate objective is to provide shepherds with rams incorporating the prolificacy of the Garole, the growth potential of the Awassi and the meat quality and hardiness of the Bannur. A very important but at present neglected application of Biotechnology which requires considerable investment is in the field of disease diagnosis and vaccine preparation. Once productive animals that respond to inputs are available and the rearers can afford to pay for the above services I am confident that the private sector will enter this field to provide this input.

It is imperative to review the cost benefit ratios of Government institutions and their contribution to the Goat and Sheep industries I think you will find that funds given to NGO's are more productive. There is much talk of helping NGO's but real assistance only seems to come from foreign aid agencies which is unfortunate.

Bacterial Biotechnology

The Production of high quality agricultural products in quantities sufficient to feed the world's people is clearly a matter of vital importance. Any general improvement in agricultural practices can have a tremendous economic impact, and the methods of biotechnology suggest at least two important strategies for bringing about widespread improvements.

Improvement through the use of symbiotic or potentially pathogenic microorganisms- Plants live in intimate relationship with many microorganisms. Leaf surfaces are covered by layers of symbiotic

& sometimes pathogenic bacterial & yeasts, and the soil close to the root area is populated by bacteria species very different from the species found in soil away from plants. There have been a number of attempts to modify symbiotic microorganisms so as to make them even more beneficial to plants.

Improvement through the production of transgenic plants- Plants now being cultivated represent the end result of many years of pain staking efforts at improvement. Traditional plant breeding programs rely on two methods; the selection of advantageous spontaneous mutations & the introduction of desirable traits from closely related species by crossbreeding.

Recombinant DNA Technology is bringing revolutionary change to these endeavous. It enables scientists to select precisely defined genes with well-characterized properties, clone them, and transfer them to a given plant. These genes can be taken from organisms unrelated to the target plant-even from bacteria or animals.

Transferring cloned genes from prokaryotes (bacteria) into higher eukaryotes, such as plants, is difficult. However, there are two species of bacteria, Agrobacterium tumefaciens & A rhizogenes, that transfer a small piece of DNA into plants as part of their normal life cycle.

Use of symbionts & Pathogenes

There are many Symbiotic bacterial associated with specific organs of various plants, a simple plant might be to modify such bacterial & then use such interactions with the plants to introduce the modified traits at the appropriate locations. This alternative approach is technically easier, because the engineering of bacterial DNA through recombinant DNA methods is now routine, whereas the manupulation of plant DNA is yet to be perfected.

(a) Protection of plants from frost damage via engineered symbiotic bacteria: One of the earliest examples of successful modification of a symbiotic bacterium was performed in Pseudomonas syringae, which is found at high concentrations on the leaves of many plants.

Many strains of this bacterium produce an ice nucleation protein that is apparently located on the surface of the bacteria cell, and the presence of this protein cause the formation of ice at temperature only little below 00 C, inflicting significant frost damage on important crop plants & so facilitating invasion of plant tissues by the bacteria.

(b) Use of nitrogen fixing bacteria to improve crop yields: All animals & plants & most bacteria depend on the availability in their environment of some form of "combined nitrogen" nitrate (No3), or such nitrogen containing organic compounds as amono acids. The huge amount of N2 that exist in the atmosphere are unavailable to the biological world except through the process of nitrogen fixation.

The biological process of nitrogen fixation does not require the consumption of fossil fuels or electricity does not produce environmental pollution. Thus increasing the extent of biological nitrogen fixation has been an important goal for biotechnology. According to one estimate, the biological process fixes times more nitrogen (24 x 10 17 tons/year) than is converted into chemical fertilizers by industrial processes. Thus even a small improvement in exploiting the biological process would have been a global impact.

He biological process of nitrogen fixation is complex & consumes a large number of ATP, molecules, because the enzymes involved in it must overcome the same huge activation energy barriers that faces the chemical synthesis of ammonia. Two enzymes are required: component I (nitrogenase) & component II (nitrogenase reductase). After component Ii is reduced by a strong biological reductant, 16 molecules of ATP are hydrolysed to accomplish the reduction of component I. Reduced component II alone is not capable of overcoming the activation energy barrier. Reduced component I finally reduced N2 to two molecules of NH3. Nitrogen fixation is strongly reductive reaction, and the enzymes involved are usually irreversibly inactivated when they are exposed to oxygen.

The species of closiridium & klebsilla fix nitrogen only under angerobic conditions. Other free living baueria, however, can fix nitrogen even under aerobic conditions. The cyanobaueria, which carry out oxygen-evoluing photosynthesis, perform nitrogen fixation only in specialized cells called heterocysts, which do not produce oxygen. Azotobactor consumes oxygen at an extremely high rate, and in doing so apparently protects its nitrogen fixing machinery. Another group of bacteria fix nitrogen only when they are in a symbiotic nitrogen fixer is Rhizobium, which invades the root tissues of teguminous plants, such as alfalfa, pea, clover and soyabean and lives in intracellular vacuoles. The vacuoles become filled with an oxygen-binding protein, leghomoglobin produced by the plants. This creates an environment low in free oxygen where Rhizobium is able to carry out nitrogen fixation.

Production of Transgenic Plants

To produce transgenic plants, the crucial steps are the introduction of the cloned genetic material into the plant cell nuclear & the facilitated integration of the cloned gene into the plant chromosome. It is interesting that the best method for doing so uses a system that already exists in nature, the system by which the plant pathogenic bacterium Agro-bacterium tumefaciens injects a portion of its plasmid DNA into plants & inserts it into the plant become.

The Agro-bacterium system is an excellent system for introducing foreign genes into the chromosomes of plants. The transfer into intact plant cells occur at high frequency, the T-DNA is usually integrated into the plant chromosomes at high frequencies without undergoing structural alterations, and the cells that have received the T-DNA can be selected easily by using antibiotic resistance such as the neomycin resistance maricer.

Finally the transgenic plants produced in this manner are quiet stable for at least several generations. Consequently almost all existing transgenic plants with potentially desirable traits have been obtained by this method.

Herbicide Resistance Plants

The advantages of making crop plants resistant to herbicides are obvious. Although there is a fear that such resistant may eventually increase the use of herbiude chemicals, there are also reasons to except that these transgenic plants will promote the use of sager, more biodegradable herbicides, perhaps in smaller amounts one example involves the herbicide glyphosate which inhibits 5-enol-pyruvylshikimate 3-phophate synthase-an enzyme involved in the biosynthesis of aromatic amino acids.

This enzyme has been purified from crop plants & sequenced and DNA Probes corresponding to its amino acid sequence have been synthesized. These probes were used to isolate CDNA for the enzyme from the CDNA library of a plant cell line known to overproduce 5-enol-pyruvylshikimate 3-phophate synthase. The CDNA was then cloned behind the strong CaMv promoter, and the promoter gene complex was introduced into plant cells via a disarmed Ti plasmid vector. The transgenic plants produce a much higher level of the target enzyme & thus are significantly more resistant to glyphosate. These results are encouraging because glyphosate has very low toxicity to animals is rapidly degraded in soil.

Insect Resistant Plants

Bacitlus thuringiensis is used in the biological control of caterpillars because its sporulating cells contain toxic proteins. The gene for one toxic protein was cloned behind promoters that are effectively expressed in plants & was introduced into plant cells via a Ti plasmid vector, thus producing plants toxic to caterpillar. The major problem with this approach has been the low level of expression of the toxin protein in plants, which is presumably related to the fact that the gene come from a bacterium.

Nevertheless, the method has produced tomato & tobacco plants that proved quiet resistant to caterpillars in field tests. Cotton however, is often attacked by insects that are more resistant to the B. thuringiensis toxin and the low level of expression of this toxin in transgenic cotton plants provided the plants with little if any protection. Recently, the coding sequence of the toxin was altered extensively to replace codons that are rarely used in plants as well as to preclude the formation of a strong secondary structure in MRNA. When cotton plants were provided with this modified gene, their production of the bacterial toxin increased 100-fold and they showed impressive resistant to common bepidopteran insects than damage unmodified plants.

Some seeds contain high concentrations of protease inhibitors, which are thought to interfere with the digestive process in insects. The cloning of cowpea trypsin inhibitor genes, for example and their transfer to tobacco plants have resulted in good resistance to a wide variety of leaf-eating insects.

Viral Resistant Plants

The method most frequency used for producing virus-resistant plants sprang from the observation that, of ten times, plants infected by nearly avirulent virus are thereafter resistant to super infection by a related, highly virulent one. Thus, deliberate infection with avirulent virus strain has been used to produce protection in crop plants. The method is not totally safe, however, because the avirulent strains many mutate to produce strains that are significantly pathogenic. Most plant viruses are positive strand RNA viruses covered by coat protein sub units. When the virus enters an injured plant cell the replication process begins, starting with the progressive uncoating of the virus from the 5-end of the RNA. Tranagenic plants, whose genomes contain introduced tobacco mosalc virus (TMV) coat protein genes and which continuosly synthesize the coat proteins, show resistance to virus infection.

The phenotype of these plants is consistent with the idea that resistance is a result of interference with the uncoating of the virus particle. Although the plant cells are resistant to infect TMV particles, they remain sensitive to the TMV RNA or to partially uncoated TMV particles.

The coat protein gene is usually put behind a strong promoter such as the 35S promoter of CaMV & is introduced into a plant's genome via the Agro-bacterium Ti system. Transgenic plants showing significant resistance to TMV alfalfa mosaic virus, cucumber mosaic virus, tobacco streak virus & tobacco rattle virus have already been produced in this manner. In a recent experiment two virus & potato virus Y, were introduced simultaneously into a commercially important potato cultivar, and one of the resulting trangenic plants provide quiet resistant to both viruses under field test conditions.

Chapter 3

Plant Breeding versus Plant Genetics

The cultivation of plants is an important aspect of today's society in many ways. Not only do plants supply us with a major food resource and flow of nutrition but they are also an important source of chemicals and other non-food products such as drugs, oils, latex, pigments, and resins. Because of the high value of plants, farmers are constantly trying to increase the yield and quality of their products by using more effective production techniques. The physical appearance of plants is determined by genetic factors that are inherited from parental crops. The natural transfer of genes can be altered using many genetic techniques such as selection or genetic engineering. One of the important goals of plant breeding is maintaining genetic diversity. This can be achieved through selection and hybridization. More recently, new techniques have evolved to allow us to grow crops artificially in culture from plant parts to produce whole plants. Modern science has allowed us to get away from the traditional method of plant breeding to more artificial asexual methods. This allows plants to be produced using individual, or groups of cells, which grow into full plants. Genetics has also allowed us to modify our crops in a more scientific way to select for traits that are more desired by society. In this article we will be discussing the differences between the more traditional techniques of plant breeding and the modern advances of plant genetics.

Plant Breeding

Plant improvement by man began many thousands of years ago. Pre-agricultural man learned that seeds put into the ground at a certain time of the year produced similar seed-producing plants. This was the beginning of domestication of plants and led to the production of the first crops. Soon after, it was discovered that some crops of the same species grew better or tasted better than others, which became the starting point of selection breeding.

Later in the nineteenth century, Charles Darwin brought to light his theory of *Natural Selection*, which he described as "survival of the fittest". It was this statement that showed how selection occurred naturally with the environment selecting which plants would grow in specific parts of the world. Selection by farmers was an extension of this process. Farmers were able to further select for plants that not only grew well, but had better taste and looking more pleasing to the eye. Later, selection was further applied to find crops that produced the highest yield with the best possible quality. In order for one to undergo selection, the breeder must have a large population of seeds to select from. This population should contain several different genetic types that are well adapted to the seasonal changes in the environmental conditions they are grown. These populations are called landraces and are generally found in regions where older plants have been growing for a long time. Natural selection has already occurred leaving only those crops that have been able to adapt to their environment. Landraces are generally found in areas where cross-breeding has been able to occur giving rise to many new genetic varieties with new genotypes constantly being introduced into the environment. Once an appropriate population is found, one can begin the process of selecting plants by removing those plants with undesired traits. This allows plants with the most desired traits to survive and give rise to a new line of crop that has been selectively chosen for its specific traits.

Hybrid Breeding

Another method of selection in plants occurs when two different pure lines are crossed giving rise to what is called a hybrid. Hybrids have the genetic basis of each of the traits of the parents. In this way, one can obtain plants with the genetic basis for many desirable traits exhibited by each of the parents. By further selection of the offspring, one can select for desired traits.

Plant Genetics

Although selection has been fairly effective over the years and is still used in some places today, it has not been able to keep up to the growing need for plants and plant products. Selection takes time and many things have to be considered, especially environmental factors that cannot always be controlled. It is for this reason that new and more effective methods of plant breeding have been discovered to give rise to an increased number of plants over a shorter period of time.

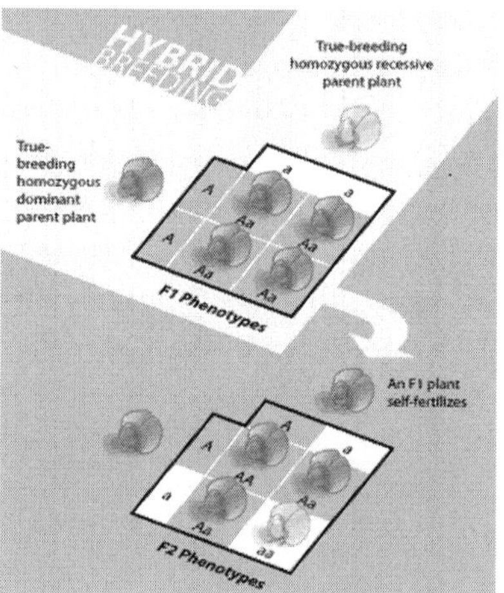

Figure: *Hybrids. Two pure breeds can by hybridized to create cross breeds which exhibit the desired phenotypes of both parent species. Here a recessive white flowered breed is crossed with a dominant pink flowered breed which contains a gene enabling it to grow in colder conditions. The desired phenotype is a white flower which can grow in the colder conditions. The offspring of the hybridization of the pure species is followed by selection and breeding to generate the desired plant.*

With a basic understanding of how genetics plays a role in determining the appearance and shape, as well as growth responses of a plant, geneticists have been able to further their knowledge and apply new techniques for growing plants.

With a foot in the door as to what gives rise to different traits, geneticists were able to take this one step further by modifying the genetics of a certain plant artificially using an asexual technique to grow these plants. This enables plant scientists to select for the traits that were most suitable for the crops they are growing without having to go through the long and tedious procedures of growing the crops and selecting for the traits that are within that population.

New techniques involve growing whole plants from single cells artificially in cultures that contain all the required nutrients and factors involved in cell growth. This technique is referred to as plant cell culture. Furthermore, scientists have been able to clone specific genes from one species and insert them into plants to introduce new traits that may benefit the plant species. This procedure is known as transgenics and gives rise to genetically modified crops. Furthermore, selection was put into operation to find crops that could withstand certain pests and environmental conditions, such as drought or freezing temperatures. An example of this is seen in genetically modified canola containing a gene that allows flounders to live in cold water. This allows the canola to survive in colder weather giving it a month longer growing season. A further example of this is seen in producing crops that are resistant to certain conditions or microorganisms that may cause harm to the plants. This also allows for farmers to reduce the amount of pesticides required in protecting their crops.

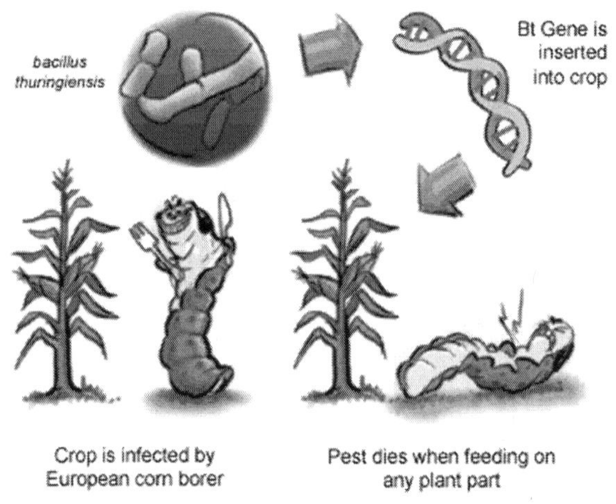

Figure: *Engineering resistant corn. Following the insertion of a gene from the bacteria Bacillus thuringiensis, corn becomes resistant to corn borer infection. This allows farmers to use less insecticides.*

New vs. Old

Although plant science is a relatively new field of interest, there has been a great deal of new technology in this field. Plant scientists have been able to overcome the slow reproductive phase involved in growing plants and have been able to produce plants that can grow in half the time with much better results. In order to make up for the growing demand of crops in today's society, working with geneticists has enabled this to occur. Although old methods of plant breeding are still practiced today in many parts of the world, especially in developing countries, many developed countries have switched to using these new technologies. However, as for anything there are limitations and concerns for both methods.

The more primitive methods of selection take a long time and do not always provide the desired crops. In addition, when a farmer does finally achieve his desired plant product, industry has changed and a new trait is preferred over the previous one. In a constantly growing and changing world, it seems that the farmer can never get ahead. Also, changing environmental conditions can lead to decreased crop yield from year to year and possibly even cause the farmer to lose entire crops. Other disasters, such as fire and disease can lend its destruction to crops causing much devastation.

In the more modern methods of plant growth, crops can be grown in artificial medium requiring less landmass to produce large amounts of crops in less time.

Although this seems like a great alternative to the earlier methods, it can also be devastating. By growing plants at a faster rate, we may be losing the essential vitamins and nutrients that are important for us. Incorporating foreign genes has been suggested to lead to allergies in many people as well as affect the food chain of other organisms. Transgenics is still a relatively new field and no concrete evidence for any of this exists but it is growing concern.

Plant science is a growing field that is constantly evolving to promote changes that benefit both the industry and accommodate the growing populations that we are facing. With the many innovative ideas and constant findings, plant science has provided many great additions to society.

Responses to Biotechnology in Crop Improvement

The advances in plant transgenics and genomics described above have not been isolated from society. Some of these achievements have been acclaimed by end-users whereas other accomplishments, e.g. release of genetically modified organisms (GMO), are being attacked, not only in words but also in deeds, by political activists.

Some of these educated middle-class campaigners are expressing in this way their rampant 'eco-paranoia', while others hide their real agenda to manipulate the fashionable ecological movement. This controversy has attracted the attention of non-scientific partizans to each side.

There have been negative comments about transgenic plants by a crown prince and contrasting positive comments by a former president, both of whom may not have the required technical knowledge to assess the potential of biotechnology for crop improvement. Irrespective of this ideological dispute and ensuing democratic disagreements, biotechnology products will be accepted by people who support scientific-based progress, in a similar way that new cultivars or innovative crop husbandry techniques have previously become integral parts of farming systems elsewhere. However, without end-user's consent, the impact of a new technology in the society will be small or nil.

Scientific honesty seems to the best policy to convince people about the advantages of biotechnology for crop improvement. What to do? Scientists, farmers, consumers, and policy-makers should objectively assess the potential hazards of crop biotechnology in farming and food systems regarding the current situation and the likelihood that such hazards may occur.

For example scientists should explain to the people that gene recombination (or reassortment) already occurs in nature. However, the ecological success of viable recombinants after gene reassortment is unpredictable owing to the high fitness of current isolates. For this reason, more scientific research will be needed to identify unpredictable risks and the chances of their occurrence.

The need for profit, as in any other business, has attracted the interest of the private sector to defend their investments in crop biotechnology with patents, intellectual property rights, and new protection methods, e.g. 'terminator' technology that inhibits germination of self-pollinated seeds. This technology protection system prevents farmers from saving seeds from their harvest for further utilization as next season planting propagules.

Three genes, each with a specific promoter, are inserted into the 'terminator' plant. One of the genes (e.g. CRE/LOX system from bacteriophages) produces a recombinase that removes a spacer between the gene producing, for example, a ribosomal inhibitor protein and its promoter such as late embryonic abundance, which only becomes active during the late stages of embryo development. This spacer with specific recognition sites blocks the gene (for the ribosomal inhibitor protein) from being activated.

Another gene (e.g. tetracycline repressor system) produces a repressor that keeps off the recombinase gene until an outside stimulus is applied to the 'terminator' plant, e.g. a chemical such as the tetracycline, or temperature and osmotic shocks.

The United States Department of Agriculture (USDA) and a cotton seed enterprise jointly acquired a patent for this concept (U.S. patent 5,723,765). Two months after this patent was announced, one of the leading agro-chemical transnationals bought the cotton seed company, although one of its officers said that it may take many years before this 'terminator gene' idea becomes a proven technology in the seed industry.

Strategic alliances, joint ventures, research partnerships, new investments, company mergers, cross-ownerships, and take-overs in the seed and agro-chemical business have also been in the news in recent months. Likewise, some leading scientists are leaving their academic appointments to join the new private enterprises in plant biotechnology. These events are happening because the private sector wants to use biotechnology to accelerate its growth in agri-business in the short-term. Nonetheless, funds to support basic and strategic research by public researchers are needed for a long-term sustainable transfer of public goods (both knowledge and technology) to the private sector or other users.

Bioinformatics

Another important factor in the successes of the genetic improvement of crops was the development of fast and more reliable computers, which allowed easier management and analysis of data as well as publication of scientific reports. The impact of the informatic revolution in crop improvement can be partially assessed by counting the number of publications indexed in Plant Breeding Abstracts (CAB International, Wallingford, Oxon, UK).

There was ca. 22-fold increase of publications in the 1930-1997 period. It was in the 1970s that indexed publications in plant breeding exceeded 10,000 per year. More publications and easy means for retrieving this information accounted for such growth of knowledge dissemination in plant genetics and breeding.

Today, rapid information exchange has been facilitated with electronic mail and access to the internet to read electronic publications such as this journal. Nowadays, information technology and DNA science are beginning to fuse into a single operation.

Computers are deciphering, and organizing the huge genetic information that may become "the raw resource of the emerging biotech economy" in the next century. Scientists working in the new field of "bioinformatics" are developing biological data banks to download the genetic information accumulated during millions of years of life evolution, and perhaps reconstruct some of the living organisms of the natural world.

Plant Genomics

This new term, defined by the development of biotechnology, refers to the investigations of whole genomes by integrating genetics with informatics and automated systems. Genomic research aims to elucidate the structure, function and evolution of past and present genomes. Some of the most dynamic fields concerning agriculture are the sequencing of plant genomes, comparative mapping across species with genetic markers, and objective assisted breeding after identifying candidate genes or chromosome regions for further manipulations. As a result of genomics, the concept of gene pools has been enlarged to include transgenes and native exotic gene pools that are becoming available through comparative analysis of plant biological repertoires. Understanding the biological traits of one species may enhance the ability to achieve high productivity or better product quality in another organism.

DNA markers and gene sequencing provides quantitative means to determine the extent of genetic diversity and to establish objective phylogenetic relationships among organisms. 'Gene chips' and transposon tagging will provide new dimensions for investigating gene expression. Molecular biologists will study not only individual genes but how circuits of interacting genes in different pathways control the spectrum of genetic diversity in any crop species. For example, more information will be available on why plant resistance genes are clustered together, or what candidate genes should be considered when manipulating quantitative trait loci (QTL) for crop improvement.

Farming in Environmentally Friendly Systems

The aims of applied plant science research for agriculture are to enhance crop yields, improve food quality, and preserve the environment where human beings and other organisms live. The best way for conservation of plant biodiversity and its environment, would be to achieve high crop productivity per unit area. In this regard, Briggs (1998) reported that as yields treble, soil erosion per ton of food decreases by two-thirds. There has been a significant yield improvement owing to enhanced crop husbandry, but in the next years progress will be achieved by changing plants that could be more suitable to sustainable and environmentally friendly farming systems. Agro-chemical corporations are developing pest and disease resistant transgenic crops to avoid pollution with pesticides in the farming system. Furthermore, food quality will become more important than crop productivity in a wealthy society. Consumers will prefer transgenic crops if they have the desired characteristics.

In the next decades meiotic-based breeding will still generate cultivars for farmers. Genetic improvement through biotechnology needs conventional breeding because (1) the elite cultivars will be the parents of the next generation of improved genotypes, (2) field testing across locations or cropping systems and over years will be needed to determine the best selections due to the genotype-by-environment interaction. As stated by Briggs (1998), "transgenes must be viewed as improvements rather than replacements for elite germplasm".

Indeed, genetic engineering may provide a means to add value by introducing synthetic or natural genes that enhance crop quality and yield, as well as protect the plant against pest and diseases. Farmers will pay more for transgenic crop propagules if they obtain extra-income after adopting biotech-derived products. For example, seeds of insect resistant transgenic crops will be more expensive than those of available cultivars but the farmer will not need to apply pesticides in their transgenic fields. Of course, patents make transgenic seeds more expensive but also farmer's benefits may be higher.

Gene Banks, DNA Banking and Virtual Plant Breeding

The sequencing of crop genomes opened new frontiers in conservation of plant biodiversity and its genetic enhancement. The advances in gene isolation and sequencing in many plant species allows to envisage that within a few years, gene-bank curators may replace their large cold stores of seeds with crop DNA sequences that will be electronically stored.

The characterization of plant genomes will ultimately create a true gene bank, which should possess a large and accessible gene inventory of today's non-characterized crop gene pools. Of course, seed banks of comprehensively investigated stocks should remain because geneticists and plant breeders, the main users of gene banks, will need this germplasm for their work. Genomics may accelerate the utilization of candidate genes available at these gene banks through transformation without barriers across plant species or other living kingdoms. Nonetheless, genetic engineering should be seen as one of the methods of plant breeding that permits the direct alteration and re-building of a crop population. "Shutting-off" genes coding for undesired characteristics may be another application of transgenics in crop improvement.

Plant breeders will change their modus operandi with the development of objective marker-assisted introgression and selection methods. Backcross breeding will be shortened by eliminating undesired chromosome segments (also known as linkage drags) of the donor parent or selecting for more chromosome regions of the recurrent parent.

Parents of elite crosses may be chosen based on a combination of DNA markers and phenotypic assessment in a selection index, such as best linear unbiased predictors. To achieve success in these endeavours, cheap, easy, decentralized, and rapid diagnostic marker procedures are required.

There are many areas of basic and strategic research in plant breeding and genetics that are being facilitated by marker-aided analysis. With molecular markers, plant biologists are reviewing crop evolution and gathering new knowledge. Such information should be incorporated into genetic enhancement programmes, especially those with an evolutionary breeding scheme. Likewise, plant ideotypes for each crop should drive the work of plant breeders. Specific plant morphotypes have been defined in rice and wheat based on accumulated knowledge of crop physiology and crop protection. The needed characteristics required to develop improved plant prototypes ensuing from such a 'virtual breeding' approach may be available in gene banks of the crop or in those of other species. Otherwise, breeders may obtain novel transgenes to develop the required ideotype.

Nowadays, the finding of new genes that add value to agricultural products seems to be very important in the private agri-business. Unique gene databases are being assembled by the industry with the massive amount of data generated by genomics research. A new term 'biosource' was coined recently to refer to a fast and effective licensed technology of pinpointing genes. With this method, a 'benign' virus infects a plant with a specific gene that allows researchers to observe directly its phenotype. Biosource replaces the standard time-consuming approach of first mapping a gene to subsequently determine its exact function. Gene identification in DNA libraries coupled with biosource technology and an enhanced ability to put genes into plants will be routine for improving crops in the next decade.

Genomics may provide a means for the elucidation of important functions that are essential for crop adaptedness. Regions of the world should be mapped by combining data of geographical information systems, crop performance, and genome characterization in each environment. In this way, plant breeders can develop new cultivars with the appropriate genes that improve fitness of the promising selections. Fine-tuning plant responses to distinct environments may enhance crop productivity. Development of cultivars with a wide range of adaptation will allow farming in marginal lands. Likewise, research advances in gene regulation, especially those processses concerning plant development patterns, will help breeders to fit genotypes in specific environments. Photoperiod insensitivity, flowering initiation, vernalization, cold acclimation, heat tolerance, host response to parasites and predators, are some of the characteristics in which advanced knowledge may be acquired by combining molecular biology, plant physiology and anatomy, crop protection, and genomics. Multidisciplinary cooperation among researchers will provide the required holistic approach to facilitate research progress in these subjects.

Pharming and Farmer-ceuticals

Growth of cities in the developed world has already replaced farmland with shopping malls, parking lots, and housing developments. Peri-urban agriculture and home gardening are also becoming very important for national food security in the developing world as a result of rapid urban expansion. Hence, new cultivars will be needed to fit into intensive production systems, which may provide the food required to satisfy urban world demands of the next century.

Specific plant architecture, tolerance to urban pollution, efficient nutrient uptake, and crop acclimatization to new substrates for growing are, among others, the plant characteristics required for this kind of agriculture. Genes controlling these characteristics may be available in gene banks for further cross breeding, which can be assisted by genomics. Peri-urban and home garden "farmers" will have to adapt to new demands from emerging urban populations with higher income. These consumers may request a more varied diet.

For example, food crops with low fats, and high in specific amino acids may be needed to satisfy people who wish to change their eating habits. If genes controlling these characteristics do not exist in a specific crop pool they may be incorporated into the breeding pool using transgenics.

Some publications anticipated that in the next millennium food will not need to be harvested from farmer's fields. Tissue culture of certain parts of the plant may provide a means to achieve success in this endeavour. For example, edible portions of fruit crops could be grown in vitro. A steady and cheap supply of these edible plant parts will be required in this new agri-business. It will take some time before such a process can be scaled-up for commercial output. Nonetheless, a patent was submitted in 1991 by a Californian biotech company for producing a vanilla extract through cell culture. Of course, this technique will not replace farming as we know it today. This biotechnique, as well as other new farming methods, offers a means for new ways of producing food, feed or fibre.

Often plants provide the raw materials for agro-industry, and not only for food or fibre processing. Active ingredients of plants have been transformed into commercial products such as medicines, solvents, dyes, and non-cooking oils for many years. Hence, it would not be surprising to see, in few years from now, entire farms without food crops but growing transgenic plants to produce new products, e.g. edible plastic from peas or plant oils to manufacture hydraulic fluids and nylon. This new rural activity may result in important changes in the national economic sector.

'Pharming' has been added to the dictionary to indicate a new kind of system to obtain medicines. For example, oral vaccines appear to be a convenient delivery system for vaccination throughout the world. Biotechnology has been used to engineer plants that contain a gene derived from a human pathogen. An antigenic protein encoded by this foreign DNA can accumulate in the resultant plant tissues. Results from pre-clinical trials showed that antigenic proteins harvested from transgenic plants were able to keep the immunogenic properties if purified. These antigenic proteins caused the production of specific antibodies in injected mice.

Mice, which ate these transgenic plant tissues, also showed also a mucosal immune response. Arakawa et al. (1998) recently demonstrated the ability of transgenic food crops to induce protective immunity in mice against a bacterial enterotoxin such as cholera toxin B subunit pentamer with affinity for GMI-ganglioside. Also, potato tubers have been used successfully as a biofactory for high-level output of a recombinant single chain antibody.

Risk Assessment of Transgenic Crops

Lack of scientific data, non-scientific partizan views, uncertainty of potential risks, and ignorance confound rational discussion concerning the release of GMO. The issue of releasing genetically modified plants (GMP) into the farming system has become particularly agitated by lobbyist groups in Europe despite widespread cultivation of such crops in North America and elsewhere. Scientists must realise that the general public are concerned that an uncautious approach to the manipulation and cultivation of transgenic crops may affect biodiversity and its sustainable utilization in the farming system, e.g. loss of variability and viability. People also want that their views about applications of biotechnology for improving agriculture are listened irrespective of their knowledge in the subject. Moreover, farmers are afraid that negative propaganda jeopardizes the public image of their products. Scientists and policy markers should not forget that people's acceptability is the most important component of the general public assessment of risk, which includes both uncertainty and negative consequences. This acceptability depends on cultural factors because people's views change according to time and location.

The process of risk assessment in agro-chemical consists of

(i) hazard identification,
(ii) exposure assessment,
(iii) effect's management,
(iv) risk characterization, and
(v) risk management.

However, transgenic crops may be able to invade (or colonize) and multiply in many habitats. Hence, this risk assessment of a genetically modified living organism (also known as GMLO) must consider other characteristics not included when assessing the release of non-living compounds to the environment, e.g. horizontal gene transfer between transgenic crops and wild related species. Scientific risk assessment of transgenic crops must be strictly performed and precautionary principles should be considered in the decision making process. In the industrialized world, this precautionary principle is a key component of the response to the unforeseen (and sometimes irreversible) human and environmental impact, which may occur by introducing into the system new advances ensuing from research and technology development. In Norway, an unique legislation advocates that "the production and use of GMO should be ethically and socially justifiable in accordance with the principle of sustainable development" as well as "safe to humans and to the environment". By applying this framework, marketing applications of GMO could be rejected if insufficient documentation regarding ecological and heath aspects was submitted by the producer.

What are the potential ecological risks associated with the release of GMP into the farming system? These are of course a very large number of potential risks, However, perhaps the two most important risks are:

(i) GMP establishes in semi-or natural habitats, and
(ii) inserted transgenes incorporate into other species, thereby affecting non-target organisms in farms or natural habitats.

Hierarchical test protocols have been proposed to assess the risks of releasing GMP. Such protocols require knowledge about evolutionary history, morphology, life-history characteristics, pollination or breeding system, gene-transfer likelihood, natural hybridization, recruitment and vegetative propagation of a chosen species. Likewise, producers should provide, to facilitate this risk assessment, additional information regarding biochemical, physiological, and morphological changes owing to inserted gene(s), along with a list and description of marker and reporter genes included in the transgenic plant. It would also be important to add details concerning when and in which plant tissues or organs will be expressed the modified function or phenotype.

Nonetheless, people must also know that scientists assessing risks of transgenic crops may extrapolate the outcome or results from simple short-time experiments into complex long-term natural-or farming systems. Investigations about gene flow and competing ability of transgenic crops may be easily addressed through short-term experiments. However, the assessment of the environmental impact of GMP requires a long-term, expensive, holistic research. Computer modelling, which integrates knowledge about gene flow, competing ability, spread of transgenes to weedy species, and cultural practices in the farming system, may provide an alternative means for long-term risk assessment of releasing GMP into the environment.

Consumer concern about transgenic crops also focuses on their safety as food, especially if modifications could influence their metabolism or health. In this regard, transgenic plants without selectable markers, such as antibiotic resistance genes, are needed to convince GMP-sceptics of the advantages of genetic engineering for crop improvement. In this way, their criticism concerning the potential risks of transgenic crops could be overcome. For example, molecular or metabolic markers may provide a means to identify transgenic plants with desired trait(s). Of course, these alternative markers should be safe from an environmental and health perspective.

Outlook

Within the next 10 or 20 years, five research areas may become very important for crop improvement:

(i) apomixis to fix hybrid vigour,

(ii) male sterility systems with transgenics for hybrid seed in self-pollinating crops,

(iii) parthenocarpy for seedless vegetables and fruit trees,

(iv) short-cycling for rapid improvement of forest and fruit trees, and

(v) converting annual into perennial crops for sustainable agricultural systems.

The development of perennial crops will be especially important to protect the soil from erosion. Plant biotechnology will play, of course, an important role in achieving research and development success in these areas.

Banning transgenic crops in the farming system will be foolish because the potential benefits are so great. Environmentalists should recall or re-read 'Silent Spring' by Rachel Carlson (1962). Whatever scientists do to develop crops that eliminate or reduce the utilization of polluting agro-chemicals in the farming systems must be welcome by farmers and consumers. For example, one interesting approach for developing resistant transgenic crops may be through the improvement of the plant's own defence system. Inducible and tissue specific promoters could assist in this endeavour.

Collective approval may lead to new partnerships, cooperation or joint ventures in research and development between scientists in the public and private sectors that will benefit farmers and consumers with profits and high quality products, respectively. Any potential risk in human development associated with biotechnology applications in agriculture will be easily resolved in a democratic society. The public need to choose between being safely self-regulated or to follow safety regulations as agreed by lawmakers after listening to the views of scientists, producers, and consumers.

The general public should see biotechnology as a safe tool for scientific crop improvement, because it helps in the fight against hunger and poverty. Therefore, research funding should be allocated accordingly to long-term plant breeding programmes, which include biotechnology as one of its tools. In this way, we may effectively face the serious challenge of feeding the rapidly growing world population in the next millennium.

Polyploidy in Plants

One of the remarkable features of living material is their ability to perpetuate themselves. However, the ever dynamic nature of the surrounding environment has imposed upon plants, much like other organisms, various evolutionary and selective bottlenecks necessitating the adoption of ways and means by organisms to keep their "race going".

A concept which gains prominence in this regard is that of hybridization and has been an area of much fascination since the eighteenth century. Stebbins (1958) defined hybridization as the "crossing between individuals belonging to separate populations which possess different adaptive norms". Polyploidy, a prime facilitator of speciation and evolution in plants and to a lesser extent in animals, is associated with intra and inter-specific hybridization. The purpose of this paper is to give a broad overview of the phenomenon of polyploidy in its entirety in plants ranging right from a brief historical background to where it stands today. Given the importance of polyploidy in speciation this will be looked into a little more in detail compared to other aspects of polyploidy.

So what is polyploidy? It refers to a definite arithmetic relationship between the chromosome numbers of related organisms. It has been defined as the possession of three or more complete sets of chromosomes and has been an important feature of chromosome evolution in many eukaryotic taxa including plants, yeasts, insects, amphibians, reptiles, fishes and even the mammalian genome. Polyploidy is present to at least to some extent in most members of the plant kingdom being more common in some and rather rare in others. The fact that it is widespread many plants is also kind of exemplified by the wide variations in chromosome numbers with chromosome numbers ranging from 2n = 4 to 500 in angiosperms to 2n=6 to 226 in monocots. Thousands of angiosperm species have 14 to 15 pairs of chromosomes.

The phenomenon of polyploidy gained much of what it is today during the early part of the twentieth century. One of the early examples of a natural polyploid was one of De Vries's original mutations of *Oenothera lamarckiana* (mutat. *gigas*). The first example of an artificial polyploid was by Winkler (1916) who in fact introduced the term polyploidy. Winkler was working on vegetative grafts and chimeras of *Solanum nigrum* and found that callus regenerating from cut surfaces of stem explants were teratploid. Digby (1912) had discovered the occurrence of a fertile type *Primula kewensis* from a sterile inter-specific hybrid through chromosome doubling but failed to realize its significance in the context of polyploidy. Though unaware of the 'Primula type" fertile hybrid, Winge (1917), from his studies on the chromosomal counts of *Chenopodium* and *Chrysanthemum* found that chromosome numbers of related species were multiples of some common basic number; he subsequently proposed a hypothesis that chromosome doubling in sterile inter-specific hybrids is a means of converting them into fertile offsprings. This was subsequently verified by various workers in artificial inter-specific hybridizations of *Nicotiana, Raphanobrassica* and *Gaeleopsis.* Finally the colchicine method of chromosome doubling was developed by Blakeslee and Avery (1937) and became an important tool for the experimental study of polyploidy. Before going ahead what is the main significance of polyploidy in brief? As pointed out at the very outset it is recognized as one of the main process in the evolutionary history of plants and to some extent other organisms.

Though differences do exist with regard to the nature of its role and the relative importance of different kinds of polyploidy, an understanding of the ways in which polyploidy operated in the past to produce new species and races may provide useful insights to improving our cultivated plants. In fact many of our crop species, including wheat, maize, sugar cane, coffee, cotton and tobacco, are polyploid, either through intentional hybridization and selective breeding (e.g. some blueberry cultivars) or as a result of a more ancient polyploidization event (e.g. maize).

Added to it, technological advances in the analysis of genome structure and function has made it possible to better analyze the genetic consequences of genome duplication. The importance of polyploidy in such diverse fields such as cytogenetics, physiology, breeding, cytotaxonomy and biogeography in conjunction with new possibilities put forth by various molecular techniques has all spurred a resurgence of interest in issues of origin and establishment of lineages.

Origin of Polyploids

There are various modes for the origin of polyploids. These mainly include mechanisms such as somatic doubling during mitosis, non-reduction in meiosis leading to the production of unreduced gametes, polyspermaty (fertilization of the egg my two male nuclei) and endoreplication(replication of the DNA but no cytokinesis). Endoreduplication however, is more similar to somatic doubling and is therefore not viewed as a separate mechanism by some authors. These mechanisms are reviewed in greater detail below.

Chromosome doubling can occur either in the zygote to produce a completely polyploid individual or locally in some apical meristems to give polyploid chimeras. Somatic polyploidy is seen in some non-meristematic plant tissues as well. (eg: tetraploid and octoploid cells in the cortex and pith *Vicia faba).* In somatic doubling the main cause is mitotic non-disjunction. This doubling may occur in purely vegetative tissues (as in root nodules of some leguminous plants) or at times in a branch that may produce flowers or in early embryos (and may therefore be carried further down). Spontaneous somatic chromosome doubling is a rare event and the only well documented instance of the same was in case of tetraploid *Primula kewensis* which arose by somatic doubling in certain flowering branches of a diploid hybrid. The phenomenon of chromosome doubling in the zygotes was best described from heat shock experiments in which young embryos were briefly exposed to high temperatures. Zygotic chromosome doubling was first proposed by Winge and the spontaneous appearance of tetraploids in *Oenothera lamarckiana* and amplidiploid hybrids in *Nicotiana* were shown to be a result of zygotic chromosome doubling.

A second major route of polyploid formation involves gametic "non-reduction" or "meiotic nuclear restitution" during microsporogenesis and megasporogenesis resulting in unreduced 2n gametes. Non reduction could be due to meiotic non-disjunction (failure of the chromosome of separate and subsequent reduction in chromosome number), failure of cell wall formation or formation of gametes by mitosis instead of meiosis.The classic example, *Raphanobrassica*, originated by a one step process of fusion of two non-reduced gametes. The production of non-reduced gametes has been shown to be rather common in *Solanum sps.*.

Another route may involve non–reduction occurring in one of the germ lines. A tetraploid individual can then result from a two-step process (sometimes referred to as a triploid bridge mechanism) from the fusion of an unreduced 2n gamete with a reduced 1n gamete to give a 3n zygote followed by the subsequent fusion of a 3n gamete with a normal 1n gamete in the next generation to give rise to a tetraploids individual (as in artificial *Galeopsis tetrahit*).

This is a more common route of polyploid formation from unreduced gametes (though the frequent sterility of most triploid hybrids has lead to a questioning of this method by some authors rather than the former one. The production of non-reduced gametes is also a function of the environment and genotype. eg: adverse growing conditions were shown to favour an increase the number of non-reduced gametes in *Gilia*. An example of the influence of genotype in modulating the production of non-reduced gametes can be seen in case of maize wherein the gene "*elongate*" on chromosome 3 was found to increase the proportion of diploid eggs produced.

Studies on unreduced gametes in both plants and animals are getting easier with the use of rapid screening techniques such as flow cytometry, chromosme painting and other genomic techniques.

Polyspermy is observed in many plants but its contribution as a mechanism for polyploid formation is rather rare except perhaps in some orchids. Endoreduplication is a form of nuclear polyploidization resulting in multiple uniform copies of chromosomes. It has been known to occur in the endosperm and the cotyledons of developing seeds, leaves and stems of bolting plants. In animals it occurs in certain tissues such as the liver cells, and megakaryocytes (the cells which give rise to the thrombocytes).

Miscellaneous Factors Promoting Polyploidy

There are a number of other factors favouring polyploidy include (but not limited to), the mode of reproduction, the mode of fertilization, the breeding system present, the growth habit of the plant, size of chromosomes etc. These are looked into in a little more detail below.

Polyploidy seems to be favoured in long lived/perennial plants possessing various vegetative means of propagation (eg: *Fragia, Rubus, Artemisia, Potamogeton* etc.) and in those with frequent occurrences of natural inter-specific hybridizations. Various possible reasons have been advanced by various workers to account for the above phenomenon; one of the widely accepted ones' being the enhanced chances of somatic doubling made possible in plants with enhanced lifespan and vegetative means of reproduction..

Cross fertilization and allogamy were argued to be factors favouring polyploidy. Autogamy however was thought to restrict it. There have been opposing views as well on the same. For example in the tribe *Midinae* (Compositae) polyploidy was found to be highly developed in the autogamous genus *Madia* but almost absent in the allogamous species of *Layia* and *Hemizonia*.

As regards latitude and altitude, the proportion of polyploids has been found to increase with latitude and altitude; but this has not always been found to be true with respect to altitude. Various reasons which have been put forth to explain the above trend in the distribution of polyploids some of them being the better adaptability of polyploids to colder climates and changes that might have taken place in the Pleistocene period etc. Various ecological factors also have a bearing on the distribution of polyploids eg: polyploids were found to be more frequently distributed in wet soils and meadows as opposed to more stable habitats with drier soils or forest communities respectively. With regard to the breeding system since the main mode of origin of allopolyploids in annuals is by the fusion of unreduced gametes, the presence of an outcrossing breeding system tends to reduce the chances of union of unreduced gametes.

A perennial growth habit tends to favour polyploidy as opposed to an annual growth habit, probably due to the fact that having a long life span increases the chances that rare events will occur (e.g. polyploidization following hybridization), and allows for mating between polyploids and their offspring. A reciprocal relationship has been observed between cell size, chromosome size and chromosome numbers in polyploids.Added to these a few other factors related to the genotype and the environment which had been looked into during the discussion on "the production of unreduced gametes" also come into play in.

Classification of Polyploids

Kihara & Ono (1926) first described two distinct types of polyploids: "autopolyploids' and "allopolyploids". Autopolyploidy is the doubling of the same chromosome set while allopolyploidy is the product of inter-specific hybridization; it is the product of doubling in a species hybrid and is therefore a polyploid containing separate sets of non-homologous chromosomes. The frequency of multivalent formation at synapsis was initially emphasized as a criterion for distinguishing auto-and allopolyploidy. Later other basis for classification were also put forth. But nevertheless the classification of a plant into an auto or allotetraploid simply based on criteria such as resemblance in external morphology to some diploids, the behaviour of chromosomes at meiosis etc. does not give any information as to the behaviour or phylogenetic origin of the species and may therefore be of little practical value. More extensive classifications were subsequently given by other workers (Simonet, Lilienfield and Clausen, Keck and Hiesey) and the classification has expanded ever since. Below is generally accepted classification of the different classes of polyploids:

Autopolyploids: Here the genomes coming together in the polyploid are identical. Here again they may be a *strict /true autoployploid* (AAAA) or an *interracial autopolyploid* (AAAA). Autopolyploids are also called polysomicpolyploids and can occur at the level of triploidy or anywhere upwards.

Amphiploid: This was a term coined by Clausen, Keck and Hiesey (1945) and includes true allopolyploids, segmental allopolyploids and autoallopolyploids and aneuploids.

A true allopolyploid or a disomic polyploid is a polyploid species derived from hybridization of parents that had structural dissimilarity between their basic genomes. These can occur at any level from tetraploidy upwards.eg:

$$AA \text{ X } BB \rightarrow AABB$$

In a segmental allopolyploid the genomes in the species are partially homologous to each other and therefore exhibit partial multivalent formation. eg:

$$A_1A_1 \text{ X } A_2A_{-2} \rightarrow A_1A_1A_2A_2$$

Autoallopolyploids are polyploids which combine the characteristics from both autopolyploids and allopolyploids, such as. eg:

$$\text{AA X BBBB} \rightarrow \text{AABBBB}$$

Aneuploids involve the gain or loss of a single chromosomes.

Paleopolyploid: This refers to an ancient polyploid that later became a diploid again due to sequence divergence between duplicated chromosomes as in the human genome. They generally have large basic chromosome numbers.

Neoploypoids: newly formed auto and allopolyploids. A polyploid could have individuals with a series of ploidy levels within the species thereby giving rise to a 'ploidy series'. The ploidy series may consist of individuals with even or odd multiples of the basic chromosome number(eg: *Chrysanhemum* (x = 9); series 2x, 4x, 6x, 8x,10x) or odd multiples of the basic chromosome number (eg: *Crepis occidentalis* (x = 11); series 2x, 3x, 4x, 5x, 7x and 8x forms).

Aneuploid series represent succession of allopolyploids based on different basic chromosome numbers. (eg: *Stipa* 2n = 22, 24, 28, 32, 36....82). Dibasic polyploids are the sum of two different diploid numbers (eg: *Brassica oleracea(2n = 18)* and *Brassica campestris (2n = 20)* and their tetraploid derivative *B. napus* (2n = 4x = 38).

Polyploids often tolerate the loss of one or more chromosome pairs which at times may give rise to modified polyploid series, what Darlington called a 'polyploid drop'(eg: a modified series found in *Hesperis* where different species have gametic numbers of n = 7, 14, 13 and 12).

Consequences/implications of Polyploidy

Polyploidy and the Evolution of New Species

Polyploidy has been regarded as a major force in evolution and speciation. It is estimated that between 47% and 70% of angiosperm species are polyploid and this shoots up to as high as 95% in Pteridophytes (reaches the highest known levels in the plant kingdom). Polyploidy is rather common in some families like Rubiaceae, Compositae, Iridaceae, Gramineae etc. It is uncommon but present in other families such as Caesalpinaceae, Passifloraceae and Fagaceae.

Homosporous pteridophytes are found to have higher base numbers as compared to heterosporous types. Again the highest percentage is found in perennial herbs and a smaller proportion is found in annuals and woody plants. It has been studied to a lesser extent in thallophytes (algae) but polyploids have been also been observed in Cladophora and Chara. A few cases of polyploidy have been observed Rhodophycophyta and Phaeophycophyta. In bryophytes mosses seems to be a classical example for the occurrence of natural and artificial polyploids but is rather uncommon in liverworts. Stebbins even pointed out to inter-generic differences in the frequency of polyploids and cited the example of *Saliaceae* where polyploidy was common in *Salix* but rare in *Populus*. In Gymnosperms it is rare. It is unkown in cycads and *Ginko* but is known in *Ephedra, Pseudolarix amabilis* and *Seqoia sempervirens*.

Inspite of the prevalence of high rates of polyploidy in many of the plant species there have been opposing views on the as well on relative contribution of polyploidy towards the process of speciation. These are exemplified in the statements below:

> *...polyploidy has contributed little to progressive evolution.*
>
> *....polyploidy, far from playing a secondary role in evolution, has provided the additional, uncommitted gene loci necessary for major steps in the evolution of animals.*

The field of polyploidy therefore formed an area of interest and controversy for more than 50 years. The application of molecular techniques is now contributing to reshape many of our traditional tenets of polyploidy which is looked into in greater detail in some of the sections below.

The evolutionary forces involved in speciation can broadly be grouped into those producing variation and those tending to fix it. The forces mainly contributing to bring about variation include hybridization and novel mutations while those tending to fix it include selection with wide outcrossing, inbreeding by self fertilization and selection, inbreeding by assortative mating etc.. The variation generating and variation fixing forces can be combined in various permutations and combinations. The process of polyploidization can essentially be seen as a variation generating force involving extensive intra and inter-specific hybridizations as mentioned earlier. Naturally occurring inter-specific hybrids, presumably F1s occur in all major groups of plants. Examples for artificially induced polyploids range from the classical *Raphanobrassica* example of Karpechenko to many of the modern crop plants. The contribution of polyploidy in the evolution of new species can be better illustrated by looking into some specific examples below.

The classic intergeneric cross between *Raphanus* (radish, 2n=18) and *Brassica* (cabbage,2n=18) caused the formation of a sterile F1 hybrid with only univalents. However subsequent chromosome doubling resulted in a fertile allotetraploid, *Raphanobrassica* (2n=36), which behaved similar to normal diploids and therefore provided a clear example of how hybridization between two species followed by chromosome doubling could cataclysmally (cataclysmic evolution) bring about the creation of a new species.Though an artificially created allopolyploid species, it definitely epitomized all criteria required of a legitimate species encountered in nature. *Primula kewensis* also represented another example of an artificial allotetraploid created from a previously sterile hybrid (got from a cross between two diploid species).

The contribution of natural polyploidy to the creation of new species can best be seen from the stories of various crop plants such as wheat (*Triticum aestivum*), tobacco (*Nicotiana tabaccum*), and upland cotton (*Gossypium hirsutum*) to name a few.

The origin of allohexaploid bread wheat (*Triticum aestivum*) occurred from three diploid grass species coupled with chromosome doubling.

The New World cottons (2n=52) are also supposed to have risen from an allopolyploidization event between Asiatic cottons (2n=26) and American cottons (2n=26).

For long polyploid events were considered rare and most of the polyploid species were thought to have single origin. Polyploids in general were regarded to have the ability to colonize a wider range of habitats and survive better in harsh and unstable climates as compared to their diploid progenitors probably due to increased heterozygosity and genic and biochemical flexibility provided by the presence of additional alleles. Due to the 'buffering effect' of multiple genomes, mutations and recombination etc. they were thought to contribute lesser to the development of new adaptive complexes than diploids.(considering equal mutation rates in diploids and polyploids). There again, allopolyploids were considered a major force in evolution while autopolyploids on the other hand were considered rare and rather maladaptive.

But as mentioned earlier many of these views on seem to be undergoing transformation largely due to data available from molecular approaches such as RFLPs, GISH, chromosome painting etc. Polyploids in fact are thought to represent a relatively frequent class of mutation, one that occasionally establishes within populations when its phenotypic effects are relatively mild. Or conversely polyploidy may be common because polyploid species may evolve faster or in more novel directions.

Most of the polyploid species are now seen to be of multiple origin with a few exceptions such *as Arachis hypogea* and salt water grass (*Spartina anglica*). The exact extent of multiple origins is unknown but in most cases is thought to be underestimated.Studies on recent allopolyploidy in *Tragopogon* indicate that multiple origins can occur frequently over a short time and area.(just past about 50 years in a small region of eastern Washington). The concept of recurrent polyploidization and subsequent further interbreeding of genotypes is best seen in the Artic flora. Multiple polyploid events from genetically and morphological differentiated diploid populations yield a complex of different genotypes at the ploidy level and these genotypes in turn ultimately come into contact and hybridize.

Another discovery is the occurrence of extensive genomic reorganization in polyploids. Chromosome painting, genetic mapping and comparative genetics have shown the occurrence of extensive intra and intergenomic reorganizations. Nine intergenomic translocations were detected in allotetraploid tobacco (*Nicotiana*). Soybean, an ancient tetraploid also revealed significant intra-chromosomal rearrangement. In another experiment by Song *et al.* using synthetic polyploids of *Brassica* to study the evolution of early generations after polyploidization, it was found that extensive genomic changes were detected in F5 as compared to the F2 plants as indicated by the loss/gain of parental restriction fragments and also by the appearance of new fragments. Transposable elements (TE) might also facilitate genome restructuring since duplicates of all genes in polyploids might buffer them from the deleterious consequences of transposition and TE will tend to multiply and be maintained.(eg: in cotton the spread of A genome repeats to D genome is thought to be mainly by transposition).

In addition to gene restructuring, in a polyploid organism there could be several fates for redundant genes including regulatory or functional divergence which will preserve the increased gene number but may also lead to 'copy number dependent gene silencing'; an epigenetic change.. Extensive silencing may finally cause a polyploid to no longer appear to be structured as an allopolyploid as in *Zea mays*. Gene silencing from epigenetic modifications is thought to be reversible.

Also it has long been held that polyploid species have the ability to colonize a wider range of habitats and survive better in harsh and unstable climates (better heat, cold drought resistance etc.. But an opposing view has been presented by Stebbins who postulated that the better ability is due to secondary contacts between previously isolated populations which generate more aggressive and adapted gene combinations that have been buffered and maintained. In a 39 year experiment with diploid and tetraploid *Ehrharta erecta* he found that the polyploid was inferior in behaviour to their diploid ancestors under field conditions.

The high rates of polyploid occurrence has been looked into and explained on different grounds. Incompatible gene or chromosome combinations that may be brought together in allopolyploid genomes cannot be easily flushed out through Mendelian segregation and the elimination of such DNA sequences and the alteration of DNA methylation patterns permit fertility restoration in some allopolyploids. Another contradicting view in the context of polyploids with respect to evolution is the fact that autoploids are not maladaptive to evolution. Important genetic attributes such as increased enzyme multiplicity and increased allelic diversity in fact make them a strong success in nature.

Other Consequences of Polyploidy

The effects of polyploidy on evolution have been one of the one key areas of focus of many workers and authors. Nevertheless polyploidy does have implications on various other aspects some of which are looked into below.

Cytology: Polyploids in general tend to have larger cells compared to their diploid progenitors. As regards the effect of polyploidy on the volume of the cell there seem to be conflicting reports with some stating the lack of any effect cell volume while others reporting an increase in cell volume (upto eight times that found in haploids). Overall there seems to be a reduction if the surface area to volume ratio which in turn has a bearing on enzymes such as ornithine transcarbymylase, tryptophan synthase, invertase and acid phosphatse related to cell volume and cell surface area respectively. Metabolism and growth seems to be rather retarded in polyploids due to altered geometric relations between the nucleus and the rest of the cell. An inverse relationship between DNA content and development rate has also been found. Polyploids also seem to have reduced number of cell divisions during development.

Gene activity: There have been reports of variable results with regard to the effect of polyploidy on gene activity, the total RNA and protein contents. Differential amplification of RNA cistrons were reported with increasing ploidy levels in *Datura innoxia* and autotetraploid tomato. With regard to effect of chromosome doubling on enzyme behaviour "ambiguous" results have been reported.The levels of various enzymes involved in the electron transport chain, photosynthesis and photorespiration were found to increase, decrease or remain unchanged in autotetraploid tomatoes. The CO2 exchange rate has generally been reported to decline with ploidy level but has been found to increase in *Festuca arundinacea.* The effects of chromosome doubling may exhibit differences between gametophytes and sporophytes of the same species and also between different strains within the same species.

Growth substances: As regards the effect of polyploidy on various growth hormones lower contents of abscisic acid and an auxin like substances were found in autotetraploid *Lycopersicon pimpinellifolium* as compared to diploids.

Water balance: A higher water content and a lower osmotic pressure was found in autotetraploid tomatoes and also in *Petunia.* Not much was know about differences in the rates of water uptake but it was found that transpiration rates were lower in polyploids.

Responses to Different Stress

Parameters such as whole plant weight, relative water content etc. decrease to smaller extent in polyploids as compared to diploids. Generally polyploids seem to be more tolerant to drought than their diploid progenitors. But contradictory views have been presented by Stebbins who observed that tetraploids of *Ehrharta erecta* persisted only on shady well drained soils whereas the diploid s survived in more harsh environments. As regards cold tolerance it was found that autotetraploids of *Brassica campestris, Raphanus sativus* and a few others were more cold resistant while tetraploids of *Trifolium repens* were less resistant. Polyploids plants were also found to be more resistant to mutagenesis and irriadiation. Polyploids seem to be more tolerant to poor soils, since most of them are slow growers and therefore have more modest nutrient requirements. Most polyploids in general are more resistant to pathogens and pests due to enhanced production of various secondary plant metabolites. Polyploids are also more resistant to herbicides, which may be due to increased heterozygosity and increased genetic redundancy.

Other Aspects

There seems to be reduced pollen and seed viability in autopolyploids as compared to diploid ancestors. There is an alteration in the primary and secondary metabolism. Autopolyploids of many drug plants have increased quantities of alkaloids per dry weight. Chromosome doubling may also alter the secondary metabolism in a qualitative manner (eg: differences in glycoflavone profiles). Another effect appears to be reduction in the amount of branching, which occurs frequently not universally. Seeds of most tetraploids are larger than those of diploids; however the percent germination may be variable being lower in some (eg: *Lycopersicum esculentum*) or higher in others (*Oryza punctata*). The leaves and other appendages of polyploids in general seem to be more thicker, shorter and broader. Polyploidy also seems to affect incompatibility relationships of some self–incompatible species. Any incompatibility in the diploid parents such as in *Raphanus* and *Brassica* seems to be carried down equally to the autotetraploid also.

A more thorough treatment of these effects can be found in 12, 13 and 26.

Polyploids in Agriculture and Horticulture

Some of the crop plants cited by Stebbins (1950), which are polyploids, include potato, coffee, banana, peanut, tobacco, wheat, oats, sugarcane, plum, loganberry and strawberry. After observing the polyploid nature of different crop plants and also the bigger size and vigour of the same in many cases, plant breeders started getting interested in the artificial induction of polyploids. Though this was initially marred by the lack of a suitable method for the induction of polyploidy, the discovery of the "colchcine method" served to remove this hurdle to a large extent. Polyploids are now found in alsike clover, rey, turnip, dill, spinach, apple, radish, grapes, zinnias, sugarbeets, petunia, *Datura, Polulus*, tea, watermelons, various forage grasses etc. to name just a few. It must bc pointed out that in many of the seed crops polyploids produced had lower fertility rates than their diploid prototypes also in general there is an optimum range of polyploidy beyond which growth may be depressed with increasing chromosome numbers. But in plants and flowers where seed cost are minor compared to the beauty of the flowers or other valuable features of other plant parts reduced fertility may not be a critical factor.

Polyploidy is an exciting phenomenon with intriguingly, a lot of practical potential. As mentioned earlier it has been an area of interest right from the eighteenth century and continues to be an area of interest as seen from the increasing number of publications being generated continually on various aspects of polyploidy over the last decade.Technological advancement and various molecular tools(such as GISH, RFLP, flow cytometry etc.) seem to be helping us gain better insights into various aspects of polyploidy in plants and animals and also help reshape many of the traditionally held tenets. While it is suggested by some that that certain factors such as the high rates of polyploid occurrence in plants, the absence of ethical issues etc. may make plants very good models to elucidate the molecular mechanisms of polyploid formation yet others have suggested a 'cross-disciplinary' approach involving both plants and animals in looking into the origin and establishment of polyploids lineages as a more efficient strategy.

Many discoveries of the past in conjunction with the current information being generated seem to be setting the stage for a new series of questions such as the extent of geneflow in polyploid lineages of separate origin, the rapidity of genome restructuring, its prevalence in neoployploids and the like.

Many of our crop plants are polyploids and a number of them have and are being synthetically produced. Yet, it has been impossible to predict with certainity, which diploid genomes when merged will coexist stably. Therefore transforming the present 'hit and miss' process into a more exact science will form an area of focus in future. Also a more detailed analysis of gene expression changes in many of the newly formed polyploids could help uncover the impact of new variation on polyploid evolution. An understanding the timing and frequency of changes in many of the newly formed polyploids along with the mechanisms involved in diploidizationmay also facilitate gain a more thorough insight into the evolutionary impact of rapid genome changes in neoployploids along with strategies possible for manipulation.

Breeding Perennial Grain Crops

Annual grain crops have dominated the earth's agricultural landscape since the time of the earliest farmers 10,000 years ago. Soil erosion followed tillage agriculture as it spread across the earth's surface. In the last few decades, one-third of the planet's arable land has been lost to soil erosion. The pace of this degradation will increase as the limits of our food-production capacity are stretched to feed a growing population of humans and domestic animals.

It may indeed be possible to expand food production to feed 10 billion people by the year 2050; however, if grain continues to be produced in 2050 by methods that erode soil and waste other nonrenewable resources, the Earth may not have the capacity to sustain adequate food production into the 22nd century. Tillage agriculture on sloping land always brings the risk of soil erosion.

The increasingly common practice of no-till production of annual crops, designed to control soil loss, has so far required increased use of herbicides. Furthermore, direct-seeded annual cropping systems have been shown to produce as high or higher nitrate emissions as tillage systems. Research on wholly new agricultural systems may provide the means to produce food on otherwise marginal lands and arrest or even reverse losses of ecological capital associated with many current systems. Our best examples to follow in developing solar-powered, less-polluting, soil-conserving forms of agriculture are natural systems.

Natural systems agriculture, still in the experimental phase, is a new approach to crop production. The principles of natural systems agriculture can be applied to any food-producing landscape, but most research to date has been conducted in central Kansas, USA, aimed at developing a "domestic prairie", with herbaceous, perennial grain-producing crops grown in polyculture. Other, less far-reaching recommendations for reducing soil erosion also involve perennial grains. But all of our major grain crops are annuals, and no current perennial species produce sufficiently high yields. Plant breeders must develop an array of perennial grain crops — grasses, legumes, and plants representing other families — before natural systems agriculture or other methods of producing food without soil erosion can succeed.

Wagoner (1990a) published in this journal a comprehensive review of efforts to develop perennial grains in the grass family. Her article made a convincing argument for development of perennial grains, provided a thorough history of breeding programs worldwide up to 1990, and recommended approaches to be taken in further research on breeding and crop production. We will first re-examine the feasibility of perennial grain crops, building on Wagoner's discussion. Then we will review genetic research and efforts to breed perennial grains in the grass family since 1990 and examine some possibilities for breeding perennial grains other than grasses. Finally, we will undertake an expanded discussion of breeding methodology relevant to all perennial grains.

We will keep the discussion general, recognizing that the term "perennial" has very different meanings in different environments. We will be discussing tropical and temperate crops, crops of tropical origin grown in temperate zones, and temperate crops that can be grown in the tropics. But we will discuss only those species on which research aimed at perennial grain production has been done or proposed, with concentration on the environment targeted by each research project.

Perennials to Annuals, and Now Back to Perennials

Prevalence of Annual Grain Crops

Whyte (1977) noted that the rapid warming of the earth's climate at the end of the Pleistocene Ice Age 11,000 years ago created three large and three small "arid cores" on the Asian continent. On the fringes of these cores there formed concentric "isoexothermic zones", where the highest annual temperatures occurred during the dry season. The new climatic regime in those belts favoured annual grasses and legumes, which could survive long hot, dry periods in the form of seeds. Their seeds were relatively large and could germinate and grow quickly with seasonal rains and moderating temperature. Annuals largely displaced perennial species in isoexothermic zones such as the Fertile Crescent of southwest Asia, becoming "suddenly and abundantly available" to human hunter-gatherers.

According to Whyte (1977), the wide availability of annuals in the various isoexothermic zones led humans, who previously had relied in part on seeds of perennial grasses for food, to initiate the agricultural revolution and carry it well beyond its regions of origin. Wagoner (1990a) summarized the subsequent events that led to the human species' almost complete reliance on annual species for use as grain crops, with tiny pockets of perennial grain production persisting into modern times. As she points out, "our ancestors took the easy route" by concentrating on annual grains.

A difficult road lies ahead for breeders intending to develop perennial grains. Breeding herbaceous perennials for adaptation in regions where the vegetation until recently was almost entirely made up of herbaceous perennials (e.g., the prairie of the central United States) should be eminently feasible, but simultaneous selection for persistence and grain yield will require intensive work. To domesticate wild perennials with no genetic input from other species would entail a genetic retreat of 10,000 years. Fortunately, it is possible to shorten this new round of crop development by orders of magnitude. Many perennial species can be hybridized with related annual crops, allowing us to incorporate genes of domestication much more quickly than did our ancestors who first selected the genes.

Resource Allocation and Negative Correlations

Breeders have before them the genetic resources for breeding perennial grains, but just beyond lies the question of a "tradeoff" between grain yield and perenniality. In the words of Wagoner (1990a), "Yield from a perennial grain will probably never be as high as that from annuals because the life strategies of annuals and perennials are so different.... The photosynthetic energy assimilated by a perennial plant over the course of a growing season must be divided among its perennating structures and seeds."

Indeed, high-yielding perennial grains do not exist today. Wagoner's (1990a) survey of 51 experimental studies in 27 species of perennial grasses showed that seed yields most often fell below 1000 kg/ha but could exceed that level. Piper (1999) reported similar yields in perennial grasses; however, two perennial legumes, Illinois bundleflower (*Desmanthus illinoensis*) and wild senna (*Cassia marilandica*), yielded up to 2000 kg/ha. Suneson et al. (1963) reported that their 25 years of work with perennial wheat (*Triticum aestivum* hybridized with perennial grasses) in California had produced lines whose first-year yields fell "within the range of the lowest yielding commercial wheat varieties" of the time, with a rapid yield decline in subsequent years. Recently, eight intergeneric wheat lines selected for regrowth ability in Washington state, USA yielded between 1600 and 5800 kg/ha, compared with almost 9000 kg/ha for the popular annual wheat cultivar 'Madsen'.

Is a tradeoff between perenniality and grain yield inevitable? The fundamental assumption of tradeoff theory — that the pool of carbon to be shared by reproductive and vegetative structures is fixed and cannot be increased by breeding— is open to question. Grain production can be sink-limited; for example, shading during development of wheat inflorescences (i.e., during determination of sink size) depressed grain yield, whereas shading during grain-filling (once sink size was fixed) had much less effect. When yield is sink-limited, more or larger reproductive structures will induce greater production of photosynthate, resulting in little competition with perennating structures. Basal or rhizome-derived tillers on a grass plant are largely self-sufficient, and their own inflorescences can supply much of the photosynthate for seed development. Perennials also may be able to maintain green tissue and continue to photosynthesize late in the growing season, after the photosynthetic tissue of annuals has senesced. For all of these reasons, there need not be a gram-for-gram tradeoff between grain and perennating structures.

Most experimental studies have addressed the question of grain yield per hectare vs. persistence over seasons only indirectly. Jackson and DeWald (1994) compared half-sib populations from a population of *Tripsacum dactyloides* segregating for a 'pistillate' mutation that causes a large increase in seed production per inflorescence.

The increased seed yield did not come at the expense of plant vigour or longevity. Piper and Kulakow (1994) found no correlation between seed and rhizome production in a population of unreplicated, winterhardy F3 plants from an interspecific cross between tetraploid *Sorghum bicolor* and *S. halapense;* however, rhizome production was all but lost in backcrosses to *S. bicolor,* the cultivated species.

Any tradeoff between yield and perenniality should occur only during establishment of the first year's crop from seed. Once perennating structures are developed, they can serve as a source of carbon for plant establishment in subsequent years. By way of comparison, Wagoner et al. (1993) computed the energetic costs of sowing and establishing a hectare of annual wheat, including all energy required to produce the seed, cultivate the land, and sow the seed.

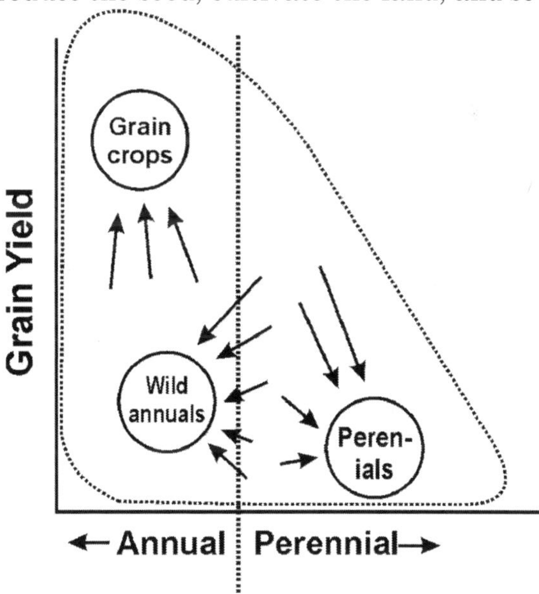

Figure: *Divergent selection pressures restricting the variability of perennial plants, wild annual plants, and annual grain crops. The horizontalal axis represents a gradient ranging from strong annuality to strong perenniality, with a threshold (dashed line) separating the growth habits. The vertical axis is grain yield per season. The dotted curve represents hypothetical genetic limits of the two-way distribution.*

The resulting cost of "annuality" was equivalent to the usable energy contained in 715 kg of wheat grain, amounting to 32% of the crop's yield in that study. Wagoner (1995) provided a table of calculations by Watt (1989) showing that intermediate wheatgrass yielding only 673 kg/ha over four years without resowing would have a break-even price of $3.60 per bushel, similar to that of spring wheat at the time.

The negative relationship assumed to exist between perenniality and grain yield is largely based on life-history theory. Gardner (1989) has provided a physiological application of the theory to crop plants. But these theories are based on observation of existing species, which are products of natural and artificial selection in divergent directions. High-seed-yielding herbaceous perennials are not found in nature — nor are triticale (X *Triticosecale*) or maize (*Zea mays*), and for the same reason. Artificial hybridization and selection produced triticale and maize, and they can, conceivably, generate productive perennial grains.

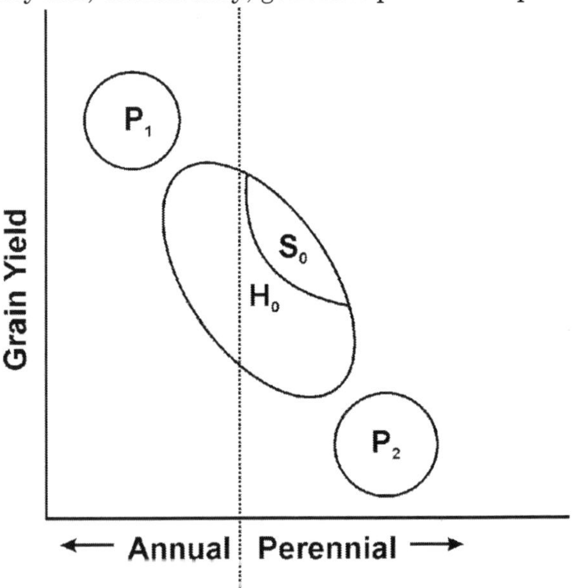

Figure: *Distributions of an annual grain-crop population P1, a perennial population P2, a population H0 derived by hybridizing P1 with P2, and a segment of H0 (S0) selected for both perenniality and grain yield.*

Even if there is a negative correlation between perenniality and grain yield, it does not preclude selection. Consider two sexually compatible gene pools: an annual population with high grain yield (P1) and a perennial one with low grain yield. If P1 and P2 are crossed, the hybrid population H0 will tend to lie in an elongated distribution between them — a "recombination spindle". The long axis is oriented between the parents because of genetic linkages and pleiotropic trait associations. The latter includes any negative correlation between grain yield and perenniality that might result from a carbon tradeoff.

Selection in the hybrid population along the long axis of its distribution would result in a larger response per generation than selection perpindicular to the long axis.

But even when selection is perpendicular to the long axis — in the direction of higher grain yield and stronger perenniality — it has the potential to approach the ancestral genetic limits or even exceed them, via new introgression or mutation. Direct selection within a perennial population for grain yield does not take advantage of genes from the annual crop and may be a longer-term project.

In any plant breeding program, negative correlations are a daily challenge. Simultaneous selection for negatively correlated traits can succeed if compromises in gains for individual traits are accepted. For example, selection for seed protein or oil concentration often has a negative effect on grain yield; nevertheless, yield and grain quality have been improved simultaneously by breeding programs.

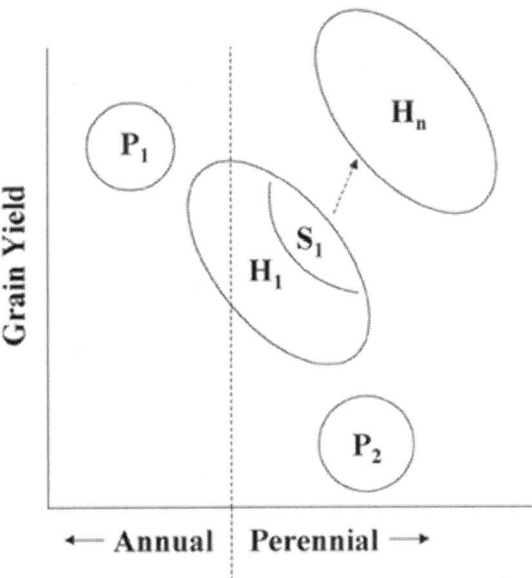

Figure: *Distributions of two parental populations, a population H1 derived by interpollinating selected individuals (S0 from Fig.1), a selected segment of H1 (S1), and a population Hn resulting from n cycles of selection for grain yield among selected individuals (dashed arrow). Direct selection within P2 for grain yield (dotted arrow) could also produce a population such as Hn.*

Plant breeders routinely make some sacrifice in grain yield to improve pest resistance, and the cost in resources and effort is real, whether or not the resistance is genetically linked to lower yielding ability. Breeders of legumes could obtain almost immediate yield increases by eliminating symbiotic fixation of atmospheric nitrogen, which has a higher energetic cost to the plant than does absorption of nitrate from the soil. But they forgo that yield increment because the biological nitrogen fixation is highly valued.

The relatively low yield of perennial grains recorded to date should not be attributed entirely to the carbon tradeoff. Perennial grains currently undergoing domestication are still essentially wild. In perennial grains being developed from interspecific hybridization, different genetic problems, each with its own array of possible solutions, have hampered breeding for yield. Among such obstacles are partial sterility caused by chromosomal differences between parental species; instability of chromosomal constitution; failure to eliminate genes for poor adaptation (unassociated with perenniality) from breeding populations; and lack of genetic diversity. If breeding perennial grains is a difficult but attainable goal, it is important for research institutions to initiate cereal-, legume-, and oilseed-breeding programs aimed exclusively at developing perennials. Today, while 33% of the world's cereal crop is fed to animals in the developed countries, we have some latitude to begin breeding crops that preserve the land while, initially at least, producing fewer total bushels than our current crops with their heavy subsidy of energy and chemicals. If we wait several decades to begin a breeding program, it may very well be too late. We will now review recent results and prospects for perennial counterparts of major annual crop species and some perennial species that have not yet been domesticated. A realistic consideration of prospects, crop by crop, may give some insight into how research resources and efforts should be allocated among and within species in coming decades.

Breeding Perennial Grains: Cool-season Grasses

Wheat

Development of perennials up to 1990: Hexaploid wheat [*Triticum aestivum,* 2n=42, genomes AABBDD] arose approximately 5000 years ago when the genomes of tetraploid wheat (*T. turgidum* or *T. carthlicum,* both 2n=28, AABB) and an Asian goatgrass (*Aegilops tauschii,* 2n=14, DD) were combined via amphiploidization (i.e., natural hybridization followed by spontaneous production of 2n gametes in the hybid.) Tetraploid wheat itself is a much older natural amphiploid, incorporating the genomes of two diploid grasses. Both wheats are part of the large and diverse Triticeae tribe of the grass family (Gramineae), which also includes scores of perennial species, many of which can be hybridized with wheat. Wagoner (1990a) examined in detail the early history of efforts in the United States, Canada, Germany, and, most importantly, the USSR, to transfer genes for perenniality from alien grass species into bread wheat, citing more than 65 publications on the subject. None of these efforts produced a truly perennial grain cultivar, but they did spin off much valuable annual germplasm with genes for disease resistances and other traits. In the end, most of the effort in the perennial-wheat programs was diverted into producing improved annual cultivars, where progress was more easily achieved.

Of the few perennial, grain-producing genotypes developed from wide hybrids at the time of Wagoner's review, none was agronomically successful. Soviet-developed 'perennial' cultivars produced good grain harvests only in the year in which they were established from seed; in the end, they were used mainly as forage cultivars that provided no more than one grain harvest. The US germplasm 'MT-2', derived from a hybrid between *T. turgidum* and *Thinopyrum intermedium* (2n=42), released by Schulz-Schaeffer and Haller (1987) in Montana, had very low kernel weight and unreliable persistence. In Sweden, Fatih (1983) found that yields of perennial *T. aestivum/Th. intermedium* partial amphiploids (2n=56) were, on average, only 48% of the yields of 42-chromosome, annual, backcross-derived lines of similar parentage.

Production of new hybrids: In the decade since Wagoner's review, no perennial wheat cultivars have been released for production. But basic research on hybridization and cytogenetics has opened up new possibilities for geneticists and breeders interested in the problem.

Species of the genus *Thinopyrum* have been hybridized with wheat more often than have any other perennial species because of the ease of producing partially fertile hybrids, often without embryo rescue. *Thinopyrum intermedium* (2n=42) is rhizomatous, and two other commonly utilized species, *Th. ponticum* (2n=70) and *Th. elongatum* (2n=14) are caespitose. Researchers at Washington State University have launched a new program to develop perennial wheat from *Thinopyrum* crosses, with promising preliminary results. As we shall, recent cytological and molecular studies have explained why *Thinopyrum* crosses have often led to frustration, and in doing so, have suggested new approaches.

Other species in the genera *Thinopyrum, Elymus,* and *Leymus* have long been investigated as sources of perenniality, and the perennial gene pool available to wheat geneticists is growing rapidly. In a review, Sharma and Gill (1983) listed only 16 perennial species that had been hybridized with hexaploid or tetraploid wheat, including only one species — *Elymus giganteus* — outside the *Agropyron-Thinopyrum* complex. A decade later, Jiang et al. (1994) added 38 additional species, including 17 in *Elymus* and 6 in *Leymus,* to the list of hybrids. Sharma (1995) reviewed the production of hybrids between wheat and more than 50 perennial species.

Our ability to make crosses and backcrosses between distantly related species has grown along with advances in embryo rescue, hormone treatments, intra-ovarian fertilization, bridge crosses, and protoplast fusion. The effects of these techniques are magnified when they are used to exploit intraspecific variation for crossability, combining ability, or variation among species carrying the same or similar genomes. Especially important are reciprocal crosses followed by embryo rescue when the traditional method of using the species with higher chromosome number fails.

Within each perennial species, accessions can vary in crossability, and choice of the annual wheat parent can also have a strong effect. For example, homozygosity for kr crossability alleles often makes hybridization possible, and may even improve early seed development; however, it may not affect results in extremely wide crosses. Choice of the wheat parent may depend on other considerations. Hybrids of *Thinopyrum* and *Leymus* with Chinese Spring (*krkr*) could not survive winter temperatures, whereas hybrids with two Japanese spring wheats were winterhardy. Fertility in the F1 through production of unreduced gametes can be induced by crossing the perennial species with the tetraploid wheat *T. carthlicum* and maintaining low temperatures during pollination and embryo development.

Many wheat/*Elymus* hybrids have been made in recent years, and addition, substitution, and translocation lines have been developed. But to our knowledge, no explicit effort to transfer perenniality from *Elymus* to wheat is underway. Lu and von Bothmer (1991) produced hybrids between 12 *Elymus* species (2n=28 or 42) and wheat (2n=42). Hybridization was made possible by using *Elymus* as the female parent and rescuing hybrid embryos, as first demonstrated by Sharma and Gill (1983). All hybrids were perennial, but chromosome pairing was very low, as expected. The hybrids were not treated with colchicine to produce amphiploids.

Hybridization between wheat and the genus *Leymus* has a long history, but has not led to perennial cultivars. Hybrids are most easily obtained with the larger-seeded, self-pollinated species of the genus: *L. arenarius* (2n=56), *L. racemosus* (2n=28), and *L. mollis* (2n=28) (Dewey, 1984), and these species also have the greatest agricultural potential. Wheat has been crossed with eight *Leymus* species in all. Most crosses require embryo rescue. Hormonal treatment has been used in producing F1 seed.

Wheat/*Leymus* hybrids produced by the Soviet perennial-wheat breeding program were perennial but less winterhardy than wheat-*Thinopyrum* hybrids. In recent studies, hybrids produced by pollinating two wheat species (*T. aestivum* and the tetraploid *T. carthlicum*) with *L. arenarius* and *L. mollis* were perennial, producing short rhizomes and exhibiting some intergenomic pairing and a high enough level of fertility to permit backcrossing. Hybrids between *T. aestivum* and two other species, *L. innovatus* and *L. multicaulis,* were non-rhizomatous.

Advances in chromosome identification: A revolution in cytogenetic techniques has led to a better understanding of the problems faced by perennial wheat breeders. MT-2 provides a good example. It was derived by selfing a 70-chromosome amphiploid containing the genomes of durum wheat (*T. turgidum,* 2n=28, genomes AABB) and *Th. intermedium,* (2n=42, genomes StStEEEStESt). Schulz-Schaeffer and Haller (1987) predicted that MT-2 would stabilize at 2n=56 through elimination of *Thinopyrum* chromosomes. In fact, according to Jones et al. (1999), individual plants of MT-2 vary in chromosome number, with most having 2n=56. But genomic *in situ* hybridization (GISH) showed that 56-chromosome MT-2 plants contained numbers of wheat chromosomes varying between 24 and 28. Therefore, there had been loss of wheat chromosomes in some plants, and *Thinopyrum* chromosomes had been eliminated at random. Plants contained up to four St-E or St-Est translocated chromosomes but no wheat-*Thinopyrum* translocations.

A wheat/*Thinopyrum* hybrid known as AT 3425, which has resistance to Cephalosporium stripe disease and perennial growth habit, has 2n=56; fluorescent genomic in situ hybridization (FGISH) and C-banding showed that AT 3425 has 36 wheat chromosomes, 14 *Thinopyrum* chromosomes, and 6 chromosomes resulting from translocations between the two species. The *Thinopyrum* chromatin in AT 3425 and another line with the same chromosome configuration, PI 550713, probably originated from *Th. ponticum,* and both lines are cytologically stable. Another perennial, 56-chromosome line, AgCs, carries the combined genomes of hexaploid wheat and the diploid species *Th. elongatum.*

Banks et al. (1992) examined meiotic pairing in crosses among eight 56-chromosome partial amphiploids derived from crosses between hexaploid wheat (*T. aestivum*) and *Th. intermedium*. As is often found, the lines had all 42 wheat chromosomes as well as 14 from *Thinopyrum*. The sets of chromosomes originating from *Th. intermedium* differed in all but two lines. Unfortunately, all eight amphiploids studied meiotically by Banks et al. (1993) were annual; two Soviet-developed perennial wheats were examined only phenotypically.

Recently, partial amphiploids that had originated in the Soviet progam have been found to be hexaploid, containing 30 chromosomes from tetraploid wheat and 12 from *L. mollis*. Thus, as in wheat-*Thinopyrum* amphiploids, elimination of the perennial parent's chromosomes had occurred. In addition, one pair of wheat chromosomes had been substituted for a pair from *L. mollis*. The partial amphiploids were annual.

Thinopyrum elongatum (2n=14) is a diploid that is more difficult to hybridize with wheat than are *Th. intermedium* or *Th. ponticum;* however, the resulting amphiploids are vigorous, stable, and perennial. Because *Th. elongatum* contributes only a single genome, all plants should have the same 56-chromosome complement. For example, the wheat/*Th. elongatum* amphiploid AgCs is cytologically stable and perennial. Jauhar (1992) produced trigeneric hybrids between a *Th. bessarabicum-Th. elongatum* amphiploid and tetraploid wheat (*T. turgidum*). The hybrids (2n=28, ABJE) were vigorous and perennial.

The maximum number of chromosomes that can be tolerated in amphiploids between wheat and *Thinopyrum* spp. appears to be 56, although 42-chromosome genotypes are more meiotically stable. Usually, partial amphiploids resulting from crosses with tetraploid wheat will contain approximately 28 chromosomes from the wheat parent and approximately 28 from *Thinopyrum,* whereas partial amphiploids with hexaploid wheat will have approximately 42 from wheat and 14 from *Thinopyrum*. Cauderon (1979) pointed out that partial amphiploids are automatically selected for "good balance" between wheat and perennial chromosomes through the elimination process; however, as we have seen, the same chromosomal complements are not consistently selected. Breeding populations based on collections of such lines would be plagued with sterility and lack of chromosome pairing unless a diploid perennial parent such as *Th. elongatum* is used.

The goal of a breeding program cannot be to develop a single perennial wheat cultivar. One partial amphiploid carefully selected to be cultivated as a perennial would have a unique chromosomal constitution. It would be a gene pool of one individual — a dead end in a breeding program. To launch a perennial wheat breeding program based on partial amphiploids would be an ambitious undertaking, involving the following steps for any polyploid perennial species targeted:

- Hybridize tetraploid and hexaploid wheats with the polyploid perennial species, making many parental combinations and sampling diversity of all parental species.
- Produce amphiploids and self-pollinate with mild selection for enough generations to achieve stable chromosome numbers.
- Use *in situ* hybridization, chromosome banding, genetic markers, and other techniques to identify chromosomes in a large population of selected plants representing many parental combinations.
- Assign partial amphiploids to groups of homogeneous chromosomal constitution.
- Compare groups for all phenotypic traits of interest and select one or a few on which to base further breeding. Develop foundation breeding pools from those groups. Experiment with intercrosses between groups, selecting for potentially superior chromosomal combinations.
- In creating new partial amphiploids to introduce into breeding pools, select strictly for appropriate chromosomal complements.

Such a plan would require a mammoth investment of resources, but the initial production of stable partial amphiploids is feasible on a large scale. Selfing and stabilization will occupy several years; by the time genotypes requiring chromosome identification can be produced, vastly more efficient cytological and molecular techniques are almost certain to be available, bringing the breeding program into the realm of the practical. But technological improvements do not guarantee the development of truly perennial grains, and parallel strategies are needed.

New strategies for perennial wheat: The perennial grasses of the tribe Triticae have long been used in wheat improvement, primarily as sources of individual resistance genes. A traditional strategy for transferring genes is to produce F1 hybrids between wheat and the donor species; either double the chromosome number of the hybrid to produce an amphiploid or pollinate the F1 directly; backcross to wheat genetic stocks to produce lines carrying the normal wheat complement of 42 chromosomes plus one or a pair of chromosomes from the donor parent; and — at some point in the process — attempt to induce a translocation that transfers a segment carrying the target gene to a wheat chromosome. There are many variations on this strategy, but the usual goal is to transfer a single gene, eliminating as much of the rest of the donor genome as possible.

Perenniality in wheat's relatives is more genetically complex than the single-gene traits transferred to date and may require a different approach. Amphiploids are not always perennial, and as we have seen, they are usually genetically unstable and agronomically undesirable. Backcrossing to wheat usually results in a return to the annual habit. Therefore, Anamthawat-Jonsson (1996) proposed backcrossing instead to the perennial parent — in their research, either *Leymus arenarius* or *L. mollis.* The objective then becomes to improve traits such as grain yield and kernel weight in the perennial species.

Anamthawat-Jonsson (1996) listed the traits to be improved by incorporating wheat germplasm into *Leymus* species: perenniality, grain quality, harvestability, threshability, kernel weight, synchronization of maturity, lodging and shattering resistance, meiotic stability, and, of course, grain yield. She has backcrossed partial amphiploids to *L. mollis,* and the progeny were vigorous with long rhizomes. In light of the many problems that have been encountered in transferring perenniality to wheat, the converse approach — using wheat to improve the perennial species — may have considerable merit. Backcrossing to the perennial will almost certainly produce breeding populations with low average grain yields and a high frequency of shattering, requiring the screening of large numbers of genotypes. Logic suggests that perennial allopolyploid species will be more tolerant of added or substituted wheat chromosomes. As we have seen, *L. arenarius* has been used as a grain crop in the past and efforts to improve it using wheat as a donor parent are underway. *Th. ponticum* is easily crossed with wheat but its high ploidy level and lack of diploidization probably would prevent its use in grain production. Wheat might be used as a donor parent for improving *Th. intermedium* or some hexaploid species of *Elymus* with which it can be crossed.

B. Rye

Diploids: Rye (*Secale cereale*) appears to be at least as promising a candidate for perennialization as is wheat. Its chromosomes are homologous with those of its direct perennial ancestor, *S. montanum.* Both species are diploid and cross-pollinated. Rye is very winterhardy, well adapted for grazing, and useful in weed control because of its allelopathic properties. The Soviet perennial-grains program included a large effort in rye, and they produced some weakly perennial genotypes that were used in limited production. Later, a decades-long effort to breed perennial rye in Germany met with only partial success. Recently, a perennial rye cultivar, 'Perenne', was released in Hungary for grain and forage production.

Despite initial expectations, no perennial rye cultivar has been used in full-scale grain production. Breeders have been stymied by the tendency of plants in *S. cereale*/*S. montanum* populations to be either fertile and annual or highly sterile and perennial. A chain of translocations involving three of rye's seven pairs of chromosomes separates the two species, and gene(s) from *S. montanum* governing perenniality are located on one or more of the translocated chromosomes. Because meiosis in plants heterozygous for one or more translocations produces many inviable gametes with duplications or deficiencies of chromosomal segments, plants in interspecific rye populations fall into one of three categories: homozygous for the *S. cereale* chromosomal arrangement (fertile, annual); heterozygous for one or more of the translocations (highly sterile); or homozygous for the *S. montanum* arrangement (perennial, fertile).

Plants in this last category would seem to answer the breeder's need; however, they are rare, and the large portion of their genomic content derived from *S. montanum* reduces their agronomic desirability and spike fertility. According to Reimann-Philipp (1995), seed-set in *S. montanum* itself is low — approximately 80%. Reimann-Philipp (1995) selected intensely for the *S. cereale* phenotype within an interspecific, perennial population. This population was presumed to be homozygous for the three *S. montanum*-derived chromosomes involved in the translocations, identified as $4R^{mon}$, $6R^{mon}$, and $7R^{mon}$ by Koller and Zeller (1976) but referred to as $2R^{mon}$, $6R^{mon}$, and $7R^{mon}$ by Reimann-Philipp (1995). He was attempting to keep these chromosomes fixed while restoring completely the other four chromosome pairs from *S. cereale* through recombination and selection. But he could not achieve a kernel weight greater than 15 mg (compared with typical values of 40 mg for annual rye under those conditions.)

As an alternative, Reimann-Philipp (1995) proposed selection for perennial plants carrying chromosomes 2R, 6R, and 7R of *S. cereale.* Presumably, such plants would arise from recombination within the ring of six translocated chromosomes. Dierks and Reimann-Philipp (1966) had postulated that perenniality was governed by a single gene that lay approximately 10 crossover units from one of the breakpoints. Selection for perenniality would be routine, but selection for the *S. cereale* chromosomal constitution would require either a laborious testcross procedure or a cytological test.

A morphological difference between chromosomes 6R and $6R^{mon}$ did not prove satisfactory for this purpose. Today, the extensive genetic map of rye could allow marker-assisted selection for the *S. cereale* arrangement. A large initial experiment could provide much more detailed information on the genetic control of perenniality; as noted by Reimann-Philipp (1995), the trait is probably affected by more than one gene.

Yet another strategy was followed by L.F. Myers and R.J. Kirchner in the breeding of 'Black Mountain' perennial rye in Australia: backcrossing the interspecific hybrid twice to the *S. montanum* parent. Perennialism (and, presumably, the *S. montanum*-type chromosomal arrangement) was quickly restored by backcrossing. But this cultivar was intended primarily as a forage grass, with *S. cereale* donating genes for nonshattering rachis and improved seed production. Oram (1996) practiced six cycles of half-sib family selection for grain and forage yield in Black Mountain, achieving gains in both traits while maintaining a low level of shattering. With grazing, stands of 'Black Mountain' decline after 3 years; however, if shattering is permitted, stands can be continually replenished by volunteer seedlings. Without selection to develop a cultivar strictly for grain production, we cannot know whether backcrossing to perennial rye while selecting for alleles from annual rye can achieve sufficient yield improvement.

Reimann-Philipp (1995) warned of a hazard when growing diploid perennial rye with the *S. montanum* chromosomal arrangement on a field scale. If pollen from the perennial drifted into seed production fields or breeding nurseries of annual rye, the resulting translocation heterozygosity would seriously and irreversibly degrade fertility in subsequent generations.

Tetraploids: In an effort to improve kernel weight, Reimann Philipp (1995) used colchicine to double the chromosome number of a perennial *S. cereale/S. montanum* population homozygous for the $4R^{mon}$, $6R^{mon}$, and $7R^{mon}$ chromosomes. The resulting tetraploid, named 'Permontra', had a kernel weight of approximately 30 mg (double that of the diploid), and first-year grain yields over 2000 kg/ha when grown in Germany. Yields declined in subsequent years. 'Permontra' achieved a similar grain yield in the Land Institute's plots in Kansas, but only 15 to 20% of the plants regrew in the next season. After a first-year harvest in the hot, dry summer of 2001, a stand of Permontra at The Land Institute died out completely by September.

Another problem with 'Permontra' — poor seed set — is also common in annual tetraploid rye, because the formation of multivalent chromosomal associations leads to production of gametes with extra or missing chromosomes. Selection can improve seed-set in tetraploid rye, and Reimann-Philipp (1995) pointed out that selection for seed set or meiotic stability can be practiced much more effectively in a perennial, by screening phenotypically or cytologically in one flowering cycle and intercrossing selected plants in the next. He found that phenotypic selection greatly improved seed-set in the perennial spring rye 'Soperta', which was derived from seven 'Permontra' plants that did not require vernalization in order to flower. Another approach would be to introduce the *Ph1* gene from wheat into a tetraploid hybrid between *S. cereale* and *S. montanum* to enforce diploid pairing and improve fertility. The *Ph1* gene was shown to operate when the chromosome carrying it was added to rye.

Tetraploid perennials may provide advantages beyond increased kernel weight. They are reproductively isolated from diploid annual seed production fields. In one study, 'Permontra' had greater heat and drought tolerance, and a much more extensive root system, than did diploid or tetraploid annuals. But perennial rye, whether tetraploid or diploid, will not be grown widely as a grain crop until the problems of sterility, persistence, and maintenance of yield over seasons are solved.

Triticale: To date, the only species to be synthesized by artificial hybridization for use as a cereal crop is triticale (X *Triticosecale*), an amphiploid of durum wheat (*T. turgidum*) and *S. cereale.* (Octoploid triticale cultivars — *T. aestivum / S. cereale*— have also been produced but not widely used commercially.) Triticale has not become one of the world's leading cereals, but its modest success suggests the possibility of developing perennial *T. turgidum / S. montanum* triticales. Derzhavin (1960b) produced and intercrossed many amphiploids derived from crosses between durum wheat and perennial rye accessions. He augmented the gene pool by allowing the amphiploids to pollinate a large number of different wheat/rye F1 hybrids. But the resulting populations were only weakly perennial. Three-way hybrids — from crosses between wheat/*Th. intermedium* hybrids and perennial rye— were more strongly perennial but had sterility, low yields, and small seeds.

Robert Metzger (USDA-ARS retired, Corvallis, OR, personal communication) reports that a *S. montanum*-derived triticale that he has developed is not sufficiently perennial, but he recommends producing and screening more new amphiploids involving a wider range of *S. montanum* germplasm. It could be that no triticale will be fully perennial, having only one of three genomes derived from a perennial species. Intercrossing of diverse lines followed by selection could improve persistence over years. If improved hexaploid or tetraploid perennial wheats can be developed, they could be crossed with *S. montanum* to produce more strongly perennial triticales.

In attempting to develop perennial triticales, breeders can take lessons from development of the annual crop. Primary triticales, i.e., newly doubled wheat/rye hybrids, inevitably suffer from sterility, seed shrivelling, lodging, and low yield potential. Decades of intense selection and introgression have resulted in triticales that are cytologically stable and improved for all of these traits, but they stand on a narrow germplasm base. Improving the performance and genetic variability of the triticale gene pool can be accomplished by several means: production of new primary triticales; triticale/wheat crosses; triticale/rye crosses; and crosses between hexaploid and tetraploid triticales.

Primary triticales derived from the wild *S. montanum* are even more agronomically primitive than *S. cereale*-based primary triticales and will require even greater breeding effort with a wide range of parents. "Substituted" triticales in which one or more wheat chromosomes replace those of rye are often agronomically superior; chromosomes 2D and 6D appear to be selectively propagated by breeders in populations segregating for R-and D-genome chromosomes. But it must be kept in mind that until genes conditioning perenniality can be mapped, random substitution of wheat for *S. montanum* chromosomes will reduce the chances of selecting a strongly perennial triticale.

Direct Domestication of the Perennial Triticeae

Intermediate wheatgrass: From domestication of perennial grasses with wheat or rye as a donor parent, it is a relatively short leap to domestication without any interspecific crossing. Three large-seeded perennial species that have been hybridized with wheat also have attracted attention as candidates for direct domestication. By far, the most work in this area has been done with *Th. intermedium* by Wagoner (1990a, 1995) at the Rodale Institute in Pennsylvania and her colleagues at the USDA-NRCS Big Flats Plant Materials Centre in New York. Wagoner (1990a) described in detail the characteristics that make intermediate wheatgrass a good candidate for domestication as a perennial grain, while noting shortcomings that must be addressed. Becker et al. (1991) concluded that its grain has protein quality "superior to the cereal grains now commonly grown", with no significant amounts of antinutrients. Recurrent selection is a logical breeding method for improving an only slightly domesticated, cross-pollinated species like *Th. intermedium*. Using mass selection without controlled pollination, Knowles (1977) increased seed yield in an intermediate wheatgrass population by 10% per cycle.

In each cycle, 1000 plants were evaluated for spike fertility over three years, and the best 50 were selected. When selected plants were removed to the greenhouse over the winter to exclude pollination by non-selected plants, thereby doubling parental control, gain per cycle increased to 20% — a result perfectly consistent with selection theory. Gridded mass selection is used to exercise control over microenvironmental effects and increase selection response. Wagoner (1990a, 1995) evaluated 300 accessions of intermediate wheatgrass, for grain yield, yield components, and end-use quality, selecting the 20 best accessions in 1989. The selections were transplanted into a polycross nursery, and 380 progeny resulting from pollination among the selections were evaluated, in a field divided into blocks of 25 plants each, between 1991 and 1994.

The best 11 plants resulting from within-and among-block selection, plus three selections resulting from further evaluation of other accessions, were put into a second-cycle polycross, and 400 individual progenies were evaluated in a second blocked nursery. Yield per plant in the 14 selections was approximately 25% higher than the population mean. Evaluation of the second-cycle population is underway, and selected plants will be intermated in 2002 to complete another breeding cycle (M. van der Grinten, USDA-NRCS, Big Flats, NY, pers. commun.) Better environmental control through selection within blocks may have produced the five-percentage-point improvement in selection response over that of Knowles (1977), but it remains to be seen if the small effective population size in the second cycle (14 plants) will restrict genetic gain in the future.

Wildrye: The Land Institute in Kansas has studied perennial cool-season grasses as potential grain crops for over 20 years. They evaluated almost 1500 accessions representing 85 species of *Agropyron, Thinopyrum, Elymus,* and *Leymus,* along with 2630 accessions of other species, between 1979 and 1987. The species selected as having the greatest potential for domestication was *L. racemosus,* known commonly as giant or mammoth wildrye. However, prospects for utilization of this species in the near future are unclear. Among 16 accessions evaluated over 2 years, yields did not exceed 830 kg/ha, and yield declined rapidly in the second and third years. Wildrye's great vigour, accompanied by large spikes but sparse seed-set, resulting in low harvest index, may provide considerable scope for breeders to select for diversion of photosynthate toward grain production. But there is no current breeding program for grain yield in *L. racemosus;* until selection is undertaken, no conclusions can be drawn regarding its potential. *L. racemosus* is self-pollinated and would require a breeding approach different from that taken with *Th. intermedium.*

Lyme grass: Lyme grass or beach wildrye (*Leymus arenarius*) has been used as a food grain since the time of the Vikings, and, as we have seen, is being studied as a potential grain crop in Iceland. There is significant genetic variation among accessions of *L. arenarius* and *L. mollis*, and it would be interesting to know which approach would result in more rapid genetic progress: direct selection within the species or an interspecific backcross program using wheat as a donor parent. The latter strategy takes advantage of genes selected through millenia of wheat domestication and breeding, but introduces chromosomal instability.

Prospects for direct domestication: Is there sufficient genetic variation within these three, or other, cool-season grasses to support large improvements in yield, kernel weight, and other traits? The very existence of annual grain crops proves that selection over thousands of years can move the mean of a species far beyond its original phenotypic range. Gains of 20 to 25% per cycle are much more rapid than typical gains in major annual crops, even considering the longer selection cycle of perennials. But to effect sufficient changes in a matter of decades rather than centuries — while possibly working "uphill" against the problem of resource allocation in perennials — will require much larger breeding efforts than have been undertaken to date. Relatively small efforts at domestication, which are within the capabilities of nonprofit organizations such as the Rodale Institute or Land Institute (or small-scale breeding programs within larger organizations such as USDA or universities), must be expanded to a much larger scale by university, government, or corporate breeding programs if wholly new perennial grain crops are to be developed.

The yield increases of 25% per cycle achieved by Wagoner et al. (1996) are remarkable, especially considering that they selected for other traits in addition to yield. But because response to recurrent selection tends either to follow a linear path or decelerate, future gains per cycle will probably be no greater than a constant percentage of the base population's yield. If a hypothetical perennial grass population yielding 500 kg/ha of grain undergoes selection, with a yield increase of 125 kg/ha/cycle (25% of the base yield), 20 cycles will be required to reach 2500 kg/ha. Because selection in perennials must be based on evaluation over two or more seasons, a single cycle can occupy four or five years. Obviously, if it is going to take almost a century to develop a high-yielding perennial crop through direct selection, a long-term commitment is required; however, such a rate of progress is much greater than the rate at which our annual crops were domesticated and improved. Marker-assisted selection and/or some genetic input from wheat could speed up the process.

Oat: Perennial oats for grain production might be developed from crosses between the cultivated hexaploid oat (*Avena sativa,* 2n=42, genomes AACCDD) and a wild, perennial, autotetraploid relative, *A. macrostachya* (2n=28, CCCC). Such crosses require embryo rescue. The F1 is highly sterile, but backcrosses to *A. sativa* have been made, and limited pairing between chromosomes from different parents does occur. J. P. Murphy has produced a 70-chromosome amphiploid between the species. The objective of this cross is to improve winterhardiness in the annual crop; perenniality has not been evaluated. Because of the partial homology that exists between chromosomes of the parental species, the amphiploid is likely to suffer from the same chromosomal instability found in wheat amphiploids. But the amphiploid, like the hybrid, can be backcrossed to *A. sativa* (Murphy, pers, commun.) Ladizinsky (1995) domesticated accessions of two wild annual oat species by using the cultivated oat A. sativa as the donor of genes for nonshattering and other traits. Perhaps this approach could be tried with *A. macrostachya.*

More hybrid combinations and larger populations will be needed if genetic studies and selection for perenniality are to succeed in the backcross generations. A more diverse sample of *A. macrostachya* parents would be desirable, but the species is restricted to two mountain ranges in Algeria, limited germplasm collections exist in the United States, and there are very few accessions held in other countries. Selection for perenniality in colder climates could be thwarted by the lack of winterhardiness in oats; winter annual oats are not generally sown above 35 degrees latitude in North America. At sites where *A. macrostachya* was collected by Guarino et al. (1991), the mean minimum temperature of the coldest month ranged from-0.6 to-3.6°C.

Rice: Although tropically adapted, rice (*Oryza sativa*) has the C3 carbon fixation pathway and is included here with the cool-season grasses. The perennial ancestor of *O. sativa* is *O. rufipogon*. Both species are diploid (2n=24) with homologous chromosomes and they can be hybridized easily. Indeed, natural hybridization and introgression occur in the field (Majumder et al., 1997). From 1995 to 2001, the International Rice Research Institute (IRRI) had a program for development of perennial rice cultivars to reduce erosion on the steep slopes where upland rice is often grown (Bennett et al., 1998; Schmit et al., 1996). Populations from IRRI's breeding program, which was discontinued in 2001, have been distributed to cooperators in China, where perennial rice breeding efforts continue.

Sacks et al. (2000) found wide variation in second-year survival among 51 *O. sativa/O. rufipogon* F1 hybrids. Sixteen percent of the hybrid combinations had greater than 50% survival, and 19% of all hybrid plants survived. In a cross between a rice cultivar with a regeneration score of 1.0 and an accession of *O. rufipogon* with a score of 3.8, the F1 had a regeneration score of 4.0, and the scores of F2 clones ranged from 0 to 5. Paradoxically, in three of the four chromosomal segments that affected regeneration ability, it was the annual parent's allele that had a positive effect. In contrast to mapped regrowth loci in *Sorghum bicolor/S. propinquum* populations, none of the regeneration loci were associated with effects on tiller number.

Selection in interspecific populations may be aided by rice's detailed molecular map and the known locations of chromosomal segments affecting traits of domestication. Many of the traits separating the annual and perennial species show polygenic inheritance. Surprisingly, *O. rufipogon* was the source of four chromosomal segments with positive effects on testcross grain yield in one set of backcrosses; however, three of the four segments were adjacent to segments that either increased plant height or delayed maturity. With positive alleles affecting perenniality *and* productivity apparently being contributed by *both* parental species, prospects for breeding high-yielding perennial rice genotypes may be bright.

Oryza sativa can also be hybridized with *O. longistaminata,* the perennial ancestor of West African rice, *O. glaberrima*. The perenniality of *O. rufipogon* lies in its ability to regrow repeatedly through production of new tillers, whereas *O. longistaminata* regrows from rhizomes. In crosses between *O. sativa* and *O. longistaminata,* genes affecting rhizome production appear to be linked to genes for hybrid embryo abortion. Consequently, IRRI scientists backcrossed rare hybrids to both parental species and intercrossing the progeny in an effort to develop a rhizomatous, agronomically acceptable genotype. Tao et al. (2001) recovered a single rhizomatous individual from among 162 plants produced by backcrossing an *O. sativa/O. longistaminata* hybrid to *O. sativa*.

A long-lived, three-species hybrid (*O. sativa/O. rufipogon/O. longistaminata*) has persisted through winters in China with monthly mean temperatures as low as 5°C. Crossing the hybrid with *O. sativa* and intermating perennial progenies has eliminated shattering. One possibility for breeding an even more cold-tolerant perennial rice exists. An ecotype of *O. rufipogon* known as 'Dongxiang' has the ability to regrow in regions of China where temperatures below -10°C are common.

Breeding Perennial Grains: Warm-season Grasses

Sorghum

Hybridization with Sorghum propinquum: In tropical environments, grain sorghum (*S. bicolor,* 2n=20) is able to regrow from basal nodes to produce a rattoon crop. But breeding a sorghum that is winterhardy in temperate regions will require transfer of genes from related species.

A perennial native of southeast Asia, *S. propinquum* is rhizomatous and diploid, with chromosomes largely homologous to those of grain sorghum. Paterson et al. (1995) evaluated rhizome-related traits of 370 F2 and 378 BC1 plants from a cross between the two species. Surviving a mild winter in southern Texas, USA, with only three nights reaching temperatures of -3°C to -4°C, 92% of F2 plants and 46% of BC1 regrew in the spring. Plants regrew either from tillers or from rhizomes. Forty-eight F2 plants representing the range of the population were selected for progeny testing.

From all F2 plants and F3 lines, Paterson et al. (1995) collected data on number of rhizomes producing above-ground shoots, distance between the centre of the crown and the most distal shoot, a subterranean rhizome score, tillering, and regrowth. They mapped chromosomal segments affecting these traits in the F2 plants and F3 lines, using 78 RFLP loci.

Rhizomatousness was a complex trait, with nine different chromosomal regions on seven of sorghum's ten chromosomes having detectable effects on at least one of the rhizome traits. Individual segments accounted for between 5 and 13% of the total variation. All but one of the seven segments associated with regrowth was also associated with one or more rhizome traits, and all four segments associated with tillering were also associated with regrowth or rhizomatousness.

Because *S. propinquum* is a tropical species, rhizomatous progeny of crosses between *S. bicolor* and *S. propinquum* would probably not be winterhardy at middle or northern latitudes, without successful selection for deeper rhizome growth. But there is great potential for developing a perennial grain sorghum for the tropics or subtropics from such populations.

Hybridization with johnsongrass: Johnsongrass (*S. halapense,* 2n=40) is a tetraploid, probably an amphiploid that combines the genomes of *S. bicolor* and *S. propinquum.* It is a very strong and aggressive perennial, and a notorious weed. Like *S. propinquum,* johnsongrass stores starch in its rhizomes. As a consequence, its rhizomes have no cold hardiness, unlike those of temperate grasses, which store fructosans. Natural selection for deeper-growing rhizomes has allowed johnsongrass to spread as a weed as far north as Ontario. Although the most northerly biotype reproduces mainly by seed, regrowth from rhizomes occurs throughout the range of the species.

Early research on hybrids between diploid sorghum and johnsongrass produced two types of hybrids: 30-chromosome plants that were male-sterile but could be backcrossed to the diploid parent and fertile 40-chromosome plants derived from unreduced female gametes in the diploid parent. The 30-chromosome plants were more strongly rhizomatous. Hadley and Mahan (1956) identified seven 20-chromosome backcross plants that were rhizomatous, but most were chlorophyll mutants. Three years of selection failed to produce a single diploid line that was rhizomatous.

Hybrids between *S. halapense* and induced tetraploid lines of *S. bicolor* are easily made. In an effort to produce a perennial grain sorghum at The Land Institute, Piper and Kulakow (1994) crossed *S. halapense* with tetraploid grain sorghum lines. In an interspecific F3 population, approximately 40% of plants were rhizomatous. There was no significant negative correlation between rhizome production and grain yield in the F3 generation, but a negative association arose with backcrossing. Yield was strongly related to plant biomass and root biomass, both with pheontypic correlations of 0.70. The F3 population — selected for winterhardiness but not for yield — had a grain yield 62% as high as the mean non-irrigated sorghum yield in Saline County, Kansas, where the experiments were conducted. But rhizome production dropped to near zero in other populations derived by backcrossing to tetraploid *S. bicolor* in an effort to increase grain yield.

Land Institute breeders have selected for winter survival among the rare rhizomatous BC2 plants and their selfed progeny. The phenotypes of winterhardy selections remain very distant from that of the cultivated parent, despite the latter's expected 87.5% genetic contribution to the BC2. The selections are taller and later maturing than either parent, have open panicles and small seed, and produce many tillers, although not as many as johnsongrass. Rhizome mass is less than 10% that of johnsongrass — sufficient for overwintering, but not enough to allow interspecific progenies to become aggressive weeds.

Both rhizomes and tillers originate from meristems at the base of the plant, and there appears to be considerable overlap in their genetic control. It may not be possible, or even desirable, to select a low-tillering, sufficiently rhizomatous genotype. Indeed, selection for yield improvement may be more effective in highly tillering populations. If a grass plant is regarded as a population of largely autotrophic tillers, and each tiller supports both seeds and rhizomes, then the most direct route to increased yield is via additional tillers. Of course, this implies increased biomass.

Piper and Kulakow (1994) concluded that development of a winterhardy sorghum (i.e., one that produces 80g of rhizomes per plant) with a grain yield of over 4000 kg/ha is feasible, through selection for greater biomass and reallocation of photosynthate to seed production. The foundation germplasm for breeding a perennial sorghum may necessarily consist of high-biomass plants that produce more tillers than annual sorghum. With a sufficiently large genetic base, subsequent selection for improved harvest index and seed size could be successful.

Because there is some homology between the chromosomes of grain sorghum and johnsongrass, multivalent chromosome associations are common at meiosis in interspecific tetraploids. Multivalents, in turn, cause poor seed set because of nondisjunction of chromosomes. Luo et al. (1992) demonstrated that selection for fertility can be effective in autotetraploid grain sorghum, which is generally plagued by low seed set. Breeders could cross perennials with the highly fertile tetraploid germplasm that Luo et al. (1992) have produced. Broadening and improving the genetic base of tetraploid perennial sorghum will require introduction of more agronomically elite germplasm. One rapid method of incorporation would be to pollinate both diploid and induced-tetraploid strains of elite, large-seeded inbred lines with the best tetraploid perennials. From the diploid/tetraploid crosses, breeders can select 40-chromosome hybrids that arise from unreduced gametes. The Land Institute is now taking this approach to develop genetically diverse breeding populations.

Pearl Millet

Pearl millet (*Pennisetum glaucum,* 2n=14), like sorghum, is a tropical, annual diploid with a perennial, tetraploid relative. Napiergrass, *P. purpureum* (2n=28) has one genome homologous and one nonhomologous to that of pearl millet. Amphiploids resulting from colchicine treatment of hybrids between the species are male and female fertile. Hanna (1990) backcrossed these hexaploids to diploid and tetraploid pearl millet lines and produced perennial progeny; however, both types of backcross plants (tetraploid and pentaploid) were highly sterile and unable to survive the mild winters of south Georgia, USA. Napiergrass, the only known species in pearl millet's secondary gene pool, is not rhizomatous, so selection for winterhardiness would probably not be successful. A perennial millet for grain production in the tropics is a reasonable prospect.

Dujardin and Hanna (1990) interpollinated hybrids and their derivatives from crosses between tetraploid pearl millet and *P. squamulatum* (2n=54), a more distant, perennial, apomictic relative. Some progenies (2n=48) were both perennial and apomictic. Apomixis can be used to ensure grain production and genetic stability in highly heterozygous progenies of interspecific crosses. *Pennisetum squamulatum* is also non-rhizomatous, and its progeny are not likely to be perennial outside of the tropics.

Maize

Hybridization with tetraploid perennial teosinte: Efforts to develop perennial maize (*Zea mays* ssp. *mays,* 2n=20) have been sporadic at best; as in other crops, hybridization between maize and perennial relatives has led primarily to improvement of the annual crop. Shaver (1964) first attempted development of maize-like perennials from crosses between colchicine-induced tetraploids of maize and a wild, perennial, tetraploid relative, *Z. mays* ssp. *perennis* (2n=40). Selection within the resulting tetraploid populations and backcrosses to tetraploid maize effectively increased the frequency of perennial progeny. Crosses to diploid maize produced perennial triploids, but all diploid selections were annual. Shaver (1967) combined a postulated gene (*pe*) for perenniality with recessive genes for indeterminacy (*id*) and grassy tillers (*gt*) in a diploid background, to produce perennial plants; however, the *idid* genotype prevented production of ears. Because

Shaver (1967) had developed a separate *idid* population in a different genetic background that did produce ears, he suggested that perennial diploids could also be made fertile if the genetic background were manipulated.

Hybridization with diploid perennial teosinte: Little further attention was paid to perennial maize until the dramatic discovery of a diploid species of perennial teosinte, *Z. mays* ssp. *diploperennis*. Initial studies showed that inheritance of perenniality was relatively simple in maize/*diploperennis* crosses, but perenniality was inferred from tillering habit, a potentially misleading technique.

In subsequent, larger-scale experiments, inheritance of tillering in progeny of similar inter-subspecific crosses was more complex, and perennial maize types were not recovered even in large segregating populations.

Genetic mapping in maize/annual teosinte crosses show that most traits of domestication separating the species are oligogenic, and the loci tend to be clustered on the map, through either linkage or pleiotropy. A similar study of these traits, plus perenniality, in crosses between maize and diploid perennial teosinte would be of great value to any breeding program attempting to combine perenniality with the agronomic phenotype of maize. This would require substantial effort to evaluate large segregating populations for tillering, rhizome production, and capacity to produce seed over multiple seasons. Once the genomic regions of interest are identified, marker-assisted selection can be used to incorporate them into a maize background and eliminate unwanted alleles such as those conditioning hard glumes and shattering.

One serious obstacle to adoption of any teosinte-derived perennial grains is the lack of winterhardiness of these tropical species. There are no winterhardy species of *Zea*. Because the bulk of maize production and breeding occurs in temperate areas, there has been little incentive to develop perennials from crosses with *Z. mays* ssp. *diploperennis*.

One possible approach has not been suggested to date: selection for rhizome depth. As we have seen, johnsongrass rhizomes also are not winterhardy if near the soil surface, but dispersal of the species into higher latitudes has been made possible by selection for deeper rhizomes. Superimposing selection for this undoubtedly complex trait on selection for perenniality and traits of domestication, not to mention yield, may entail a much larger effort than any breeding program is willing to undertake.

Hybridization with eastern gamagrass: The closest winterhardy relatives of maize are in the genus *Tripsacum*. Eastern gamagrass (*T. dactyloides*), for example, is currently grown as a perennial forage grass as far north in the western hemisphere as Kansas and Massachusetts, and can be grown in the Corn Belt. *T. dactyloides* has been hybridized many times with maize, beginning with the work of Manglesdorf and Reeves (1931). Plants of the diploid (2n=36) or tetraploid (2n=72) races may be crossed with maize. If *Tripsacum* is used to pollinate maize, embryo rescue is necessary, but if maize is used as the male, some hybrid seed may be obtained without rescue. In addition, several strains of popcorn, when pollinated with tetraploid *T. dactyloides,* produce large amounts of hybrid seed that does not require embryo rescue. Some have good crossability with diploid *T. dactyloides* as well. Contrary to typical results, Eubanks (1995, 1997) reported that a putative 20-chromosome hybrid between *T. dactyloides* and *Z. diploperennis* showed 93 to 98% pollen fertility.

Natural introgression between *Tripsacum* and maize has not been observed, but morphological and molecular evidence supports the hypothesis that the species *T. andersonni* is an intergeneric hybrid containing three genomes (54 chromosomes) from *Tripsacum* and 10 chromosomes from *Zea* in *Tripsacum* cytoplasm. The uniformity of this ancient natural hybrid indicates that *T. andersonii* arose from a single hybridization. It has been able to spread across tropical Latin America because of its vigorous perenniality.

In addition to being perennial, tetraploid *T. dactyloides* is a facultative apomict. Perennial hybrids result from artificial crosses between tetraploid *Tripsacum* and maize, and some seed-set can result from apomixis. The hybrids, derived from parents with different basic chromosome numbers and chromosomes of different sizes (those of maize being larger), are male sterile, with cytological behaviour that is anything but regular. Harlan and deWet (1977) summarized methods for utilizing such hybrids in maize improvement. Either 28-chromosome or 46-chromosome hybrids — derived from diploid and tetraploid *T. dactyloides* parents, respectively — can be backcrossed to maize. In either case, *Tripsacum* chromosomes are eliminated with backcrossing. Elimination occurs more gradually in progeny of 46-chromosome hybrids, and the 20 chromosomes of the resulting backcross plants can contain significant genetic material from *Tripsacum*. All 20-chromosome backcross plants derived to date have been annual and non-apomictic. Kindiger et al. (1996) derived an annual, 39-chromosome line that carried 9 *Tripsacum* chromosomes and displayed an intermediate level of apomixis. Most hybridization with *Tripsacum* has been for the purpose of either elucidating the evolution of maize or transferring resistance or other genes to annual maize. The latter purpose implies backcrossing to maize. But development of perennial populations may require interpollinating plants in early backcross generations that still carry many *Tripsacum* chromosomes, or even backcrossing to *Tripsacum*. As Harlan and deWet (1977) commented, "Apparently, if one wishes to contaminate maize with *Tripsacum* one should first contaminate *Tripsacum* with maize."

New approaches to perennial maize are being explored. An anomalous fertile hybrid between diploid *T. dactyloides* and maize was discovered by one of the authors (BEZ) in 1997 near the mouth of the Big Nemaha river in Richardson County, Nebraska, USA. This derivative of natural introgression between *T. dactyloides* and a putative commercial hybrid is being hybridized with gynomonoecious *Tripsacum* — both diploid and tetraploid — and with tassel-seed popcorn to develop a 56-chromosome perennial cultivar for production of grain, forage, fibre, and fuel.

The difficulties encountered in introgressing apomixis from *Tripsacum* into maize (Kindiger et al., 1996) should temper hopes for a rapid synthesis of perenniality with high grain yield. Whatever the initial population, and even with marker-assisted selection, the process of recovering perennial, winterhardy segregants with maize-like ears, and then breeding for yield and other agronomic traits will be long and arduous.

Direct Domestication of Warm-season Grasses

Eastern gamagrass: Could *T. dactyloides* be domesticated directly, without introgression of genes from maize? To do so would be an accomplishment parallel to that of domesticating maize from annual teosinte — a feat requiring thousands of years and producing genetic and physiological changes much greater than those involved in domestication of Asian cereals such as wheat and rice (Iltis, 2000; Beadle, 1980). To develop a crop from eastern gamagrass using the knowledge and techniques provided by 21st-century genetics, while leaving aside the important genes of domestication available in maize, would be an ambitious project.

Wagoner (1990a) described in detail the status of eastern gamagrass as a potential grain crop, and the species' most discouraging characteristic: very low seed yield. Interest had been stimulated by the discovery of a gynomonoecious, or pistillate, mutant (DeWald and

Dayton, 1985) in which pistillate and perfect spikelets replace the staminate spikelets of the normal inflorescence. The result is an increase of up to 20-fold in the number of seeds produced per plant; however, the seeds are small, so that the weight of seed produced per plant is increased by only a factor of 3 (Jackson and Jackson, 1999).

Plants of eastern gamagrass are large, vigorous and widely adapted. The increase in sink size made possible by the pistillate mutant may provide an opportunity to increase seed yield dramatically through increased harvest index— the yield component usually found to have had the greatest effect on yield improvement in traditional grain crops (Evans, 1998).

Furthermore, Jackson and Dewald (1994) found that the increased seed yield of pistillate plants did not come at the expense of plant vigour or longevity. Carbohydrate reserves were significantly higher in pistillate than in normal genotypes. For breeders, there is a huge pool of genetic variability available in the species (Newell and deWet, 1974; Wright et al., 1983). Although tetraploid *T. dactyloides* reproduces apomictically, parental combinations can be produced via BIII hybrids, in which an unreduced egg is fertilized by a haploid sperm (Kindiger and Dewald, 1994). Alternatively, obligately sexual tetraploid plants can be produced via colchcine treatment of diploid *T. dactyloides* plants, all of which are sexual. The genus *Tripsacum* contains many species that lack winterhardiness but have desirable traits that, potentially, could be transferred to eastern gamagrass: synchronous flowering, large spikelet number, higher seed yield, and other variations in plant morphology.

The maximum seed yield of eastern gamagrass in plots at The Land Institute has been 240 kg/ha, in the third year after sowing (Piper, 1999). It remains to be seen how rapidly yield can be improved through selection within pistillate populations. And there is another question: would the 10-, 20-, or 30-fold yield improvements required to make eastern gamagrass a viable grain crop have larger negative effects on plant vigour and persistence than did the three-fold yield boost brought about by the pistillate mutation? That increase was large relative to the grain yield of a normal plant but required diversion of only a small amount of photosynthate, relative to the plant's large biomass.

The food quality of Eastern gamagrass is excellent (Bargman, 1989). But, even if yield can be improved, other problems must be solved. One problem is disease. Infection by maize dwarf mosaic virus B has been very serious in plots at The Land Institute (Seifers et al., 1993). Also, the hard fruitcase of *Tripsacum* weighs almost three times as much as the seed itself and makes processing difficult. It was an extremely rare mutation in annual teosinte that freed the kernel from the fruitcase and allowed its use as a grain and the development of maize (Iltis, 2000). This may be a gene that breeders will be forced to transfer from maize.

Indian ricegrass: Grain is currently being harvested from a perennial grass for human food in northeastern Montana, USA. Indian ricegrass (*Oryzopsis hymenoides*), cultivar Rimrock (Jones et al., 1998), has reduced seed-shattering (Jones and Nielson, 1991) and produces gluten-free grain.

The grain is being produced, milled, and marketed under the trade name Montina. Yields vary between 250 and 500 kg/ha, but improvement through breeding may be feasible. Germplasm collections exhibit great phenotypic diversity, and very large-seeded genotypes are known (T.A. Jones, USDA-ARS, Logan Utah, pers. commun.). Certainly, other perennial grasses native to the western USA could be considered for domestication as grain producers, but no attempts have been made.

V. Breeding perennial grain legumes: prospects

Soybean

The genus *Glycine* is divided into two subgenera, *Glycine* and *Soja*. The cultivated soybean, *Glycine max,* and its close relative and ancestor *Glycine soja,* both annual diploids (2n=40; genome GG), make up the subgenus *Soja*. The subgenus *Glycine* contains 16 perennial species (Singh and Hymowitz, 1999). Numerous attempts have been made to cross perennials with *G. max,* but hybrids have been produced only with *G. tomentella* (2n=80; genomes DDEE; Bodanese-Zanettini et al., 1996; Hymowitz and Singh, 1987; Hymowitz et al., 1998; Newell and Hymowitz, 1982; Shoemaker et al., 1990; Singh and Hymowitz, 1999).

In most cases, successful hybridization has required the use of ovule culture (Newell and Hymowitz, 1982). All hybrids have been sterile, but when their chromosome number has been doubled with colchicine to produce amphiploids with the genome constitution DDEEGG, some fertility has been restored (Shoemaker et al., 1990; Singh et al., 1993).

Recently, Singh et al. (1998) produced 22 monosomic addition lines of soybean, each carrying a single chromosome from *G. tomentella*. During development of the lines, a slight tendency toward perenniality persisted up to the BC2, but all monosomic lines had an annual growth habit (R. Singh, Univ. of Illinois, personal communication). Little or no pairing occurs between chromosomes of the two species (Singh et al., 1998), so even if genes governing perenniality can be identified, their transfer to soybean will be difficult.

Perhaps even more difficult would be the incorporation of winterhardiness, which does not exist in any species of *Glycine*. Winterhardiness would need to derive from another legume species such as alfalfa, via transgenesis.

Illinois Bundleflower

To dispel doubts that herbaceous perennials can produce large amounts of seed, one need only point to a legume species of the North American prairie. Illinois bundleflower (*Desmanthus illinoensis*) has a wide geographic range, stretching well beyond the prairie, from Colorado to Minnesota to Florida to Texas (Latting, 1961). It is a good nitrogen-fixer, a preferred forage for livestock, and produces relatively high yields of large seeds with favourable nutritional profiles (Kulakow, 1999).

In evaluating a germplasm nursery of 141 highly diverse *D. illinoensis* accessions, Kulakow (1999) identified 15 in which seed shattering was reduced or eliminated, a prerequisite for grain crop. The highest-yielding 20 accessions in the first sowing (1988) produced a mean of 1500 kg/ha in their first year and 1180 kg/ha in their second year of growth. These accessions significantly exceeded the nursery's mean yield in a second experiment sown in 1990, averaging 1090 kg/ ha. Severe drought occurred in their second year of growth, and their yields fell below 500 kg/ha. In related studies, yields have approached 2000 kg/ha (Piper, 1993).

The germplasm collection studied by Kulakow (1999), along with more extensive plant collection of *D. illinoensis* and its southern relative *D. leptobolus,* can provide the foundation for breeding Illinois bundleflower as a grain crop. The chief hurdle to be overcome is the palatability of the seed. Although the seed contains no toxic levels of oxalates, cyanides, nitrates, or alkaloids, and cooked seed has digestibility and a protein efficiency ratio similar to that of cooked oats (Kulakow et al., 1990), unprocessed bundleflower seeds have a foul odour and taste. Nothing is known about the compound(s) involved.

The Land Institute is investigating induced mutagenesis, but some method of processing, as is necessary for utilization of soybeans as human food, for example, may solve the problem. Our results (DLV) indicate that roasting whole seeds may eliminate the undesirable taste.

Other Legumes

Wild senna, *Cassia marilandica,* has one of the highest seed yields of any perennial species yet evaluated by The Land Institute (Piper, 1992, 1993), but yields decline after the first year. Perhaps most importantly, the species is not known to establish associations with *Rhizobium* for symbiotic nitrogen fixation; therefore, its usefulness would be limited in cropping systems that include little or no application of inorganic nitrogen fertilizers.

The chickpea (*Cicer arietinum*) is one of nine annual species in the genus *Cicer,* which also contains 34 perennial species (van der Maesen, 1987). The perennial species are little studied; accessions representing 12 species are maintained at the Western Region Plant Introduction Station in Pullman, Washington. Two species, *C. anatolicum* and *C. songaricum,* have survived for 10 years in the field and continue to produce seed.

One accession of *C. songaricum* is white-flowered, has some degree of shattering resistance, and produces seed similar to the commercial 'desi' type (F.J. Muehlbauer, WRPIS, personal communication). The presence of these traits suggests that *C. songaricum* was once cultivated in Asia. Many species of the Leguminosae may be grown as perennials for seed production in the tropics. For example, second-crop seed yields of pigeonpea (*Cajanus cajan*) were higher than first-crop yields in two studies in India (Newaj et al., 1996; Nimbole, 1997).

Breeding Perennial Composites: Prospects

Sunflower

The relatives of the annual cultivated sunflower (*Helianthus annuus,* a diploid with 2n=34) are genetically diverse, consisting of 14 annual and 36 perennial species (Seiler and Riesenberg, 1997). If attempts to develop a perennial sunflower fail, it will not be because the available gene pool is too small. According to data compiled by Jan (1997), along with a recent study by Sukno et al. (1999), 20 perennial species have been hybridized with *H. annuus.* In approximately half of the crosses, F1 plants were perennial; sometimes one accession of a species produced perennial hybrids, while another produced annuals. The majority of hybrids have enough fertility to be backcrossed.

Some hybrids were produced before the development of the first effective embryo-rescue technique for sunflower by Chandler and Beard (1983). Heiser and Smith (1964) crossed wild *H. annuus* with the *H. decapetalus, H. hirsutus,* and *H. strumosus* (all 2n=68), and Jerusalem artichoke (*H. tuberosus,* 2n=102). They found, as have subsequent researchers, that crosses were far more successful when the polyploid species was used as the female. Only the hybrids with *H. strumosus* and *H. tuberosus* were winterhardy. Whelan (1978) failed to produce interspecific crosses with cultivated sunflower, but his hybrids between wild *H. annuus* (as the male) and two diploid perennial species, *H. maximiliani* and *H. giganteus,* were fertile enough to backcross to a cultivar. Neither the hybrids nor the backcrosses were perennial. Whelan (1978) found that the perennials differed cytologically from the annual in three translocations and a paracentric inversion. Hybrids were highly sterile, but backcrossing to *H. annuus* rapidly restored male and female fertility (Whelan and Dorrell, 1980).

More recently, many crosses with perennial diploids, tetraploids, and hexaploids have been made — usually, but not always, via embryo rescue (Georgieva-Todorova, 1984; Krauter et al., 1991; Atlagic et al., 1995; Espinasse et al., 1995; Sukno et al., 1999). However, few backcrosses have been attempted. Seiler (1991, 1993) released BC1-derived germplasms from crosses between cultivated sunflower and three perennial species: *H. hirsutus, H. resinosus* (2n=102), and *H. tuberosus.* All germplasms were annual, but hybridizing them with perennial parents could increase the probability of recovering perenniality in another round of backcrossing.

Protoplast fusion is an efficient method of producing interspecific amphiploids in *Helianthus.* Krasnyanski and Menczel (1995) produced *H. annuus* + *H. giganteus* hybrids that had good fertility but were not perennial. Henn et al. (1998) produced large numbers of *H. annuus* + *H. giganteus* and *H. annuus* + *H. maximiliani* hybrids that developed rhizomes from which shoots emerged. Colchicine-induced or protoplast fusion-derived amphiploids of different parentage could be intercrossed to develop populations for breeding perennial, grain-producing plants.

Little is known about the genetics of perenniality in sunflower, and there have been no efforts to develop a perennial crop. When cultivated *H. annuus* is used as the female parent, the F1 is usually a weak, short-lived perennial, and perenniality is quickly lost upon backcrossing to the annual. When the wild perennial is used as the female, some hybrids are more strongly perennial and bear a closer resemblance to the wild parent. Recovery of full perenniality and winterhardiness may require backcrossing to the wild parent (G. Seiler, personal communication). As in sorghum, a "wilder" plant type, with a greater number of heads per plant, may not be incompatible with good grain yield, if maturation is relatively synchronous. Breeders aiming to develop a perennial sunflower should select perennial parents based on past success in hybridization and occurrence of perennial progeny, as well as regions of adaptation. The Land Institute is investigating *H. maximiliani* as a parent because of its wide adaptation across much of the North American prairie, its vigorous growth and seed production, and its crossability with *H. annuus.*

Direct Domestication of Perennial Helianthus

Annual sunflower is the only species of *Helianthus* that has been domesticated as a seed crop. But *Helianthus* contains a wide diversity of perennial species (Seiler and Riesenberg, 1997), many of which are potential domesticates. The Land Institute has studied the common prairie species *H. maximiliani* as a candidate for direct domestication (Jackson and Jackson, 1999). It is a vigorous prairie plant with small but edible seed. Seed yield estimates have varied widely (Jackson and Jackson, 1999; Piper, 1999), and yields tend to decline after the first year of propagation in monoculture. Maximilian sunflower has strong allelopathic properties, making it potentially useful in suppressing annual weeds during the establishment of perennial polycultures (Piper, 1999). It spreads so aggressively that thinning or partial tilling to reduce stands can actually increase grain yield (Jackson and Jackson, 1999).

Establishing Gene Pools

Widening Genetic Bottlenecks

Developing the first perennial grain genotype with acceptable yield will be a demanding process in itself, but that will be only the beginning; no new agricultural system can be based on only a handful of genotypes per species. Sown over large regions, even polycultures will be genetically vulnerable if they do not incorporate intraspecific as well as interspecific variation.

Furthermore, breeding of perennial grains will quickly reach a dead end without sufficient genetic diversity. During the domestication process, most crop species suffered a similar "founder effect" that restricted the gene pool on which early cultivators drew (Ladizinsky, 1985). In probably the most extreme example, all modern bread wheat may have descended from one or a few natural interspecific hybrids that occurred around 5000 years ago (Cox, 1998). If new perennial crops are to be developed on a shorter time scale, a founder effect must be avoided.

To ensure intraspecific diversity in the field and foster further genetic improvement, breeders will need to develop deep gene pools for each crop. In rye, rice, sorghum, intermediate wheatgrass, wildrye, lymegrass, eastern gamagrass, Indian ricegrass, Illinois bundleflower, or Maximilian sunflower, this will be a straightforward matter of incorporating new annual or perennial parents, manipulating ploidy or ensuring homozygosity of translocations, and alternate rounds of selection and recombination. It will involve considerable effort, but large gene pools exist to be drawn on.

Where the perennial crop has resulted from hybridization between species with nonhomologous chromosomes — as has been accomplished in a limited way in wheat and maize and remains to be done in oat, soybean, and sunflower— introducing sufficient diversity will be a more daunting task. In such a situation, new germplasm can be brought into the breeding pool only by repeated interspecific hybridization. The human-made species triticale is an example of an annual crop developed through interspecific hybridization that has always been hampered by a restricted gene pool (Skovmand et al., 1984).

Because of their more readily available gene pools, we may expect perennial versions of rye, rice, or sorghum to be the first to move beyond the experimental stage. Creating a gene pool of interspecific hybrids in wheat, oat, maize, or sunflower may take as long as or longer than domesticating wild species such as Illinois bundleflower, Maximilian sunflower, or those of the perennial Triticeae "from scratch".

Biotechnology

We have seen that sophisticated chromosomal manipulation will be necessary in developing perennial versions of some crops from interspecific hybrids. We have not discussed genetic engineering in its narrower sense: asexual insertion of individual genes. We can speculate on the potential for transforming annual into perennial plants by gene insertion, but with the state of knowledge today, we can go no further.

No research to date suggests that perenniality is governed by a single gene, or even two or three genes, in any crop or crop relative. Although some wheat amphiploids with as little as 25 to 30% of their genome derived from the perennial parent are themselves perennial (Cai et al., 2001), we have seen that in rye, triticale, sorghum, maize, soybean, and sunflower, dilution of the perennial-derived genome to below approximately 50% often eliminates perenniality.

This, along with the lack of success by breeders of any crop to backcross a gene or chromosome conditioning perenniality into any annual genotype attests to, but does not prove, the complexity of the trait.

In a few taxa — for example, *Sorghum, Secale,* and *Oryza* — segregation of mostly homologous chromosomes can be observed in crosses between annual and perennial species. As we have seen in sorghum and rice, molecular-marker studies do not show that perenniality is simply inherited (Paterson, 1995; Kohm et al., 1997). The postulation of a single gene for perenniality on one of the three *S. montanum* chromosomes $4R^{mon}$, $6R^{mon}$, or $7R^{mon}$ (Dierks and Reimann-Philipp, 1966) has not been confirmed, and the relatively weak perenniality of the cultivar Permontra, which carries all three chromosomes, indicates that the trait is more complex.

Of course, it is not impossible that a gene could be isolated that conditions the perennial growth habit when transferred to an annual plant. But if a "perenniality gene" is identified in a particular species and cloned, its effect when transferred to any but very closely related species is entirely unpredictable.

Anamthawat-Jonsson (1996) lists other obstacles to employing gene-transfer technology to improve physiologically complex traits in genetically complex species such as the polyploid Triticeae. Transgenic technology may be useful, once perennial grain crops have been developed, in improving their pest resistance, food quality, or other more simply inherited traits; however, other breeding and cultural methods will also be available in most cases.

Breeding Perennial vs. Annual Grains

General Considerations

Almost all crop breeding involves direct selection for plant productivity — grain yield in the present context — whether the plants are annual or perennial. Discussion of differences between breeding strategies for annuals and perennials have usually concentrated on allocation of photosynthetic resources between seeds and vegetative structures (Gardner, 1989). But perennials will also require arrays of adaptive traits very different from those usually addressed by breeders of annuals.

Breeders developing perennial grains for existing monoculture systems will be faced with finding genetic solutions to problems that are exacerbated when individual plants must survive and produce over a period of years rather than months in the same patch of soil.

Breeding for resistance to diseases, insects, and adverse soil conditions will probably be even more important in developing perennial grains for monoculture than they are for annuals. But the majority of perennial grain breeders will not be selecting for adaptation to monoculture. Most efforts to breed perennial grains have environmental protection as an explicit goal. This will lead breeders to select genotypes adapted to systems like organic farming (Wagoner, 1990b; Scheinost, 2001) or natural systems agriculture (Jackson and Jackson, 1999) that receive lower subsidies of nonrenewable energy and synthetic chemicals.

Such systems usually incorporate inter-and intraspecific diversity. The greater biodiversity and better soil conditions inherent in these systems is designed to reduce pressures on the breeder to select for resistance to pests and soil problems.

Differences in Selection Criteria Between Perennial and Annual Grains

Obviously, persistence and maintenance of grain yield over seasons has been difficult to achieve in the past and will occupy much of the attention of perennial-grain breeders. We will need to consider adopting some of the methodologies used by breeders of forages (Sleper, 1987) or tree crops (Libby, 1992). Breeding of perennials obviously entails longer selection cycles, but perenniality can also have a positive effect on genetic gain, through greater control over pollination (Knowles, 1977; Reimann-Philipp, 1995), the capacity for asexual propagation in breeding nurseries, and a longer time scale for evaluating traits and genetic markers.

Avoidance of tillage will leave crop debris on the soil surface, favouring pathogens that overwinter on plant parts. The potential for damage to a perennial crop rooted in place for several years in the midst of its own debris is obvious. Breeders may need to incorporate a higher level of genetic resistance to saprophytic pathogens in perennial grains, just as they must in some annual crops intended for no-till production (Bockus and Shroyer, 1998).

Unless they are grown in more ecologically sound farming systems than are most annual grains, perennial grains may need increased resistance or tolerance to soil-borne fungi, nematodes, and viruses. Annual crops are often vulnerable to these organisms, but an infection that is late or slow to develop may not cause serious damage, and a new, initially healthy crop can be sown the next year. Crop rotation can suppress such problems in annual systems. In contrast, perennial crop plants, once infected, could be subject to damage over a period of years or killed outright. Without ecological or genetic protection, replanting and rotation cycles would have to be shortened.

Breeding for perenniality through interspecific hybridization may bring some genetic protection against pests. The majority of disease-resistance genes in wheat, to take one important species, have been transferred from other species (Cox, 1998). The extensive hybridization with wheat's perennial relatives envisioned herein would bring in a bonus of new resistance genes. The same may occur in sunflower (Seiler, 1992) and other species.

The need for rapid stand establishment from seed, especially under cooler conditions, has become even more important for breeders of annual crops with the widespread adoption of no-till agriculture (Crosson, 1981). Breeders of perennials will have to devote some attention to seedling establishment, but they can divert much of that effort into selection for persistence over seasons. Because sowing would be necessary only once every few years, much more effort could be put into using cultural methods to ensure good establishment.

Problems Rendered Less Severe by Growing Perennials in Biologically Diverse Systems

Plant breeders continually face the effects of genotype-environment interactions, with "environment" including everything except the crop populations under selection. Breeding methodology traditionally attempts to control all environmental factors except the one or few under study. In more genetically diverse systems with lower external subsidies, genotype-environment and genotype-genotype interactions cannot be controlled or eliminated but must be relied upon to drive the system.

Biodiversity, which can be manifested at many different levels in agricultural systems (Cox and Wood, 1999), is the plant breeder's friend when it provides relief from problems that would otherwise be handled genetically. Breeders of annual crops for monoculture are being asked, more and more, to find genetic remedies for problems that do not exist in natural ecosystems. Natural systems agriculture seeks to mimic those ecosystems and eliminate many problems from the breeder's checklist: soil compaction, phosphorus deficiency, low-pH and/or aluminium toxicity, and other potential consequences of annual monoculture.

Breeding programs across the globe probably spend a larger proportion of their time and effort on incorporating genes for resistance to diseases and insects than on any other activity. But lower levels of genetic resistance to diseases should be acceptable in breeding perennials to be grown in mixtures. Species diversity can provide protection against pathogens (Browning, 1974) and insects (Altieri and Nicholls, 1999). To take a simple example, a bacterial leaf spot causes severe attacks in monocultures of Illinois bundleflower, but is insignificant in polyculture because a grassy understory cushions raindrops and reduces splashing (Jackson and Jackson, 1999). Perennial crop cultivars should also be designed as populations of genotypes rather than inbred lines or F1 hybrids. Intraspecific diversity can retard development of virulent insect biotypes (Cox and Hatchett, 1986; Gould, 1986) and reduce infection by aerial, splash-borne, or insect-vectored pathogens (Garrett and Mundt, 1999, 2000; Zhu et al., 2000). On the other hand, diversity may not compensate for the increased vulnerability of perennial grains to soil-borne pathogens.

Perennial mixtures will require new breeding methodologies, as do annual intercropping systems (Francis, 1990). Other crop species, weeds, mycorrhizal fungi, nitrogen fixers, other soil microorganisms, dead organic material, and a wide variety of herbivores and their own predators or parasites will affect the growth and productivity of the crops under selection.

Conclusion

We have described parental germplasm and possible methodologies for breeding at least 20 species of perennial grains. Prospects for success vary among species, but we can be certain that highly productive perennial grains will not be developed quickly or without immense effort. For any one of these species, the debate over feasibility of breeding a perennial grain crop could be the subject of an entire book or symposium; however, the only way to answer the question of feasibility is to carry out breeding programs with adequate resources and appropriate methodologies on a sufficient time scale. A massive program for breeding perennial grains could be funded by diversion of a relatively small fraction of the world's agricultural research budget.

Chapter 4

Major Breeding Objectives

Rice Research Proposal

Rice research at RES in 2007 will continue toward the primary objective of developing improved rice varieties for California. The search, interview, and selection process for a medium grain breeder to overlap and transition with Dr. Johnson's anticipated retirement in 2008 is continuing, as it is to fill the vacant short grain breeding position. We are planning to fill both positions this season. Considerable time and effort is being devoted to recruitment of staff.

Project leaders will concentrate efforts on developing rice varieties for the traditional medium, short, and long-grain market classes. Research efforts will continue to improve and develop specialty rice such as waxy (mochi or sweet) rice, aromatic rice, and others as an adjunct breeding effort. Major breeding emphasis will continue on improving grain quality, yield and disease resistance. Efforts will be made to effectively use new as well as proven breeding, genetic, and analytical techniques. RES staff will be working to expand DNA marker screening capabilities. Following are the major research areas of the RES Rice Breeding Program planned for short, medium, and long-grain types in 2007.

Quality

Efforts to identify, select, and improve culinary and milling quality in all grain types will continue to receive major emphasis. Improved techniques for cooking evaluations are being used and screening for cooking quality expanded. The RES quality lab is supporting quality evaluation and research for variety development.

Resistance to Disease

The RES Rice Breeding Program is continuing efforts to improve disease resistance in our California varieties. Evaluation and screening for stem rot and sheath spot resistance will be conducted by the plant pathologist on segregating populations, advanced breeding lines, and current varieties. Rice blast disease presents an additional threat to California. Research and breeding activities to address rice blast have been implemented and greenhouse screening for resistance is continuing. M-208, an improved medium grain with resistance to blast race IG-1, was released in 2006 and efforts to develop improved blast resistant varieties will continue.

The Pathology Project is proceeding forward on large scale backcrossing efforts to transfer disease resistance into selected varieties, primarily medium grain. Marker-aided selection will be a part of this effort as will the use of new sources of resistance. New resistant sources and foreign germplasm will continue to be evaluated as potential parental material. Foreign germplasm will be introduced through quarantine for use in breeding and research.

Yield

Yield is a complex character that results from the combination of many agronomic traits. Emphasis will continue on breeding varieties with high grain yield potential, minimal straw for high yield, and more stable yields while maintaining and/or improving grain quality.

Tolerance to Low Temperature

Tolerance to low temperature remains an essential character needed at seedling and reproductive stage in California rice varieties. Segregating populations and advanced experimental lines will continue to be screened in the San Joaquin nursery for resistance to blanking, normal vegetative growth, minimum delay in maturity, and uniform grain maturity. Selection at UCD may be discontinued due to concerns about adjacent UC research activities. Expanded large plot yield testing is being considered at the San Joaquin nursery site. Cold tolerance data will include two seeding dates of advanced material at RES, UCCE Statewide Yield Tests, refrigerated greenhouse tests, and data from the UCD, San Joaquin, and Hawaii nurseries.

Lodging and Maturity

Improved lodging resistance will receive continued emphasis in all stages of variety development. Efforts will continue to develop improved varieties that have a range of maturity dates with major emphasis placed on early, very early rice, synchronous heading, and uniformity of ripening.

Seedling Vigour

Selection and evaluation for seedling vigour will continue on all breeding material.

Cooperative Projects

Cooperative research by the rice breeding program staff with USDA, UC, RiceCAP and others in the area of biotechnology, genetics, quality, agronomy, entomology, plant pathology, and weed control will be continued in 2007. Emphasis will be placed on applied research and more basic studies that may contribute to variety improvement.

Rice Research Priorities and Areas of Breeding Research

General Rice Research Objectives of Rice Experiment Station

The primary research objective of RES is development of high yielding and quality rice varieties of all grain types (short, medium, long) and market classes to enhance marketing potential, reduce cost, and increase profitability of rice. Rice breeding research priorities at RES can be divided into general priorities, that are applicable to all rice varieties developed for California, and specific priorities, that may differ between grain types, market classes, special purpose types, and the special interests of the plant breeding team members.

A secondary but important objective is to support and enhance UC and USDA rice research through cooperative projects and by providing land, water, and input resources for weed control, insect, disease, and other disciplinary research.

- General Rice Breeding Priorities Applicable to All Public California Rice Varieties
- High and stable yield potential
- Cold tolerance
- Lodging resistance
- Resistance to blast, stem rot, and aggregate sheath spot diseases
- Seedling vigour
- Early maturity
- Synchronous heading and maturity
- Improved head rice milling yields
- High quality rice consistent with grain type, market class, or special use

- Develop and utilize DNA marker assisted selection
- Specific Rice Breeding Priorities by Grain Type, Market Class, and Special Use
- Short Grains and Premium Quality Medium Grains
- Develop premium quality short-grain Japanese type rice varieties
- Improve premium quality M-401 type medium grains
- Improve California short grains
- Improve waxy (sweet) rice varieties
- Improvement of low amylose rice
- Develop bold grain Arborio type rice
- Rice water weevil resistance
- Calrose Type Medium-Grains
- Improve conventional medium grains
- Improve blast resistant medium grains
- Increase genetic diversity
- Utilize DNA markers for blast resistance genes with USDA researchers
- Evaluate deep water germplasm
- Long Grains
- Superior quality for table and processing
- Improve head rice milling yields and fissuring resistance
- Improve basmati types
- Develop Jasmine types
- Develop aromatic types
- Improve cold tolerance
- Improve SR and blast resistance
- Rice Pathology
- Screening and evaluation of advanced breeding lines for blast, stem rot, sheath spot, and bakanae.
- Facilitate transfer of stem rot and aggregate sheath spot disease resistance from wild species of rice and disease resistance genes identified in RiceCAP
- Mapping of stem rot resistance genes and marker aided selection for stem rot and blast in conjunction with USDA Rice Geneticist and UCD researchers.

Facilitate transfer of wide spectrum blast resistance genes to adapted medium grains using accelerated backcrossing, screening, and selection for resistance.

NCVDevolution of Plant Breeding

A major objective in modern plant breeding is the making of crop varieties with the highest possible yield potential. Yield potential is defined as "the yield of a crop when growth is not limited by water or nutrients, pests, diseases, or weeds". For farmers, whose crops are indeed limited by such constraints, this yield concept may not be seen as the most relevant objective. Relieving crops of all sorts of environmental stress, leaves light and temperature, together with varietal characteristics, as the only determinants of yield. That achieved, the same high yielding variety can be used all over a climatic zone, and target area for the variety is enormous. If stress can not be eliminated, however, adaptation to a usually site-specific environment becomes necessary. In that case, the target area for each variety will be small.

Do these different perspectives warrant different approaches to plant breeding? Conventional wisdom says no. In experiments where varieties are tested at various levels of inputs, the same high yielding variety usually comes out as top yielder at all levels. Therefore, plant breeders often claim that their varieties not only perform excellently under high-input conditions, but would also be better than traditional varieties in a low-input environment. However, when these trials are taken outside of experimental farms and the varieties are tested under local or farm conditions, researchers discover what is called 'crossover in performance'. At a certain level of stress there is a crossover point beyond which local varieties or landraces perform better than the high yielding varieties.

Farmers who experience such situations are not a tiny minority. To some degree, most if not all farmers have to cope with local stress conditions. The stress-free environment is hard to achieve, and even harder to sustain. In many favourable areas, farmers abandon the high-input technology for economic reasons or because of ecological problems, thus increasing the need for locally-adapted germplasm. But modern plant breeding, whether public or private, can not supply adapted germplasm everywhere. Only a system of local seed selection can ensure that. And that means devolution of plant breeding. Can such decentralized breeding be compatible with development needs in a changing world and meet economic aspirations in a poor society? If the answer to these questions is going to be 'yes', the decentralized breeding must be able to take advantage of the power of science as well as of the capacities of local communities. Three aspects should be considered and made mutually compatible: breeding technology, participatory research methods, and organization at community level.

Technology: Toward an Evolutionary Plant Breeding

Commercial varieties enter the market through a system of trials and official release. That system requires uniform and stable varieties, and the breeding work must be streamlined for such end results. But local selection is not constrained by those requirements and therefore enjoys some freedom that does not exist within the formal system. Farmer-breeders are free to distribute heterogeneous varieties and can allow crops to continue evolving.

Exploiting heterogeneity and crop evolution in farmers' fields are outside the scope of most plant breeding research. One exceptional experiment, however, has shed some scientific light on the issue. It was started at the University of California (UC) in 1928. Composite cross populations of barley were produced, some of which were extremely diverse in origin of sources. These populations were exposed to continuous natural selection in current modern farming environments and became the subject of studies during the career span of several generations of UC professors. It appears that after low yields in initial years, the composite cross populations gradually improved in performance and eventually became quite good yielders, with excellent yield stability and disease resistance. These results inspired Suneson (1956) to propose an evolutionary plant breeding method. After assessment of later generations of the same material, Soliman and Allard (1991) concluded that such evolutionary breeding "is unwarranted" if yield potential is the major goal. However, if disease resistance and yield stability are two major objectives, "the composite cross approach is an efficient method". This amounts to saying that a major part of world agriculture, many high input systems included, could be well served by this approach.

In short, this means constructing a body of broadly diversified germplasm and exposing it to natural selection in areas of contemplated use. For those who are familiar with traditional farmers' breeding, this may sound like reinventing the wheel. In fact it is an improvement of the old wheel of plant breeding. The first step, constructing a body of of broadly diversified germplasm, is not all that straightforward. Science has access to world collections and information sources that are unavailable to farmers. A research institute can chose relevant germplasm and make composite cross populations with an evolutionary potential, most probably far beyond that of locally available varieties.

The immediate outcome, the early generation composite cross population, will be unadapted everywhere and is likely to yield poorly. With time, however, recombinations and natural sorting will improve the adaptation, and, according to the Californian experience, narrow the gap with commercial varieties. The long term outcome could be populations that outperform commercial varieties in disease resistance and yield stability and that may be used as a source of artificial selection for high yield.

The disease resistance appears to have evolved through the building up of polygenic complexes. Therefore, it provides a durable resistance as opposed to the monogenic and, therefore, mostly non-durable resistance usually bred into commercial pure line varieties. The stability, at least to some degree, depends on the buffering effect of crop heterogeneity. The Californian experiment shows that when the population is propagated in isolation for a very long time, diversity will start declining, resulting eventually in reduced stability.

Taking the lessons from these experimental findings into the context of current development needs, a few conclusions can be drawn. We need breeding populations with a very high evolutionary potential, and these populations must be exposed to the stress conditions of, or similar to, current farm environments. Furthermore, a certain level of diversity within populations must be maintained in order to sustain evolutionary potential and yield stability. Landraces are usually found to be heterogeneous. But do they have the evolutionary potential required by current development needs? Are they being subjected to a selection pressure that ensures yield stability, and are they as high yielding as they could be?

Scientific evidence may not be available to give direct answers to such questions. Community visits may not be helpful either. A confusing picture of different selection practices and frequent change of seeds (and therefore lack of persistent long term selection), appears in many communities where traditional seed systems prevail. It is also common to see the coexistence of modern and traditional varieties. More systematic efforts and a certain level of organization are necessary to make full use of knowledge already existing within a culture, to exploit fully the potential within the locally available germplasm, and to take advantage of opportunities provided by science.

Community Organization

Widely diverse forms of organizations dealing with seed management have sprouted up at the community level in recent years. I will present an Ethiopian case representing a traditional society, and a Philippino case representing a modern society. Criteria of classifying these societies as traditional and modern are access to external markets and farm inputs. These were absent in the Ethiopian case and present in the Philippino context.

The Ethiopian case is from Tigray, in the middle of the famous Abyssinian gene centre. Renowned for genetic richness and for knowledge and culture related to seed management, the area should be expected to provide an excellent site for community seed banks. But seed banks were organized more because of poor seeds than because of genetic wealth. Poor seeds were seen as one of the reasons for poor agricultural performance. This was in the 1980s and the area was isolated by war. No external support was possible and community leaders had to look for sources of improvement within their own communities. They knew that those sources existed as, in all societies, there were experts known for their skills in traditional seed selection who had fine local seeds. But they needed an organization to extend the benefits of those experts and those seeds to the wider community.

The Philippine case is from Mindanao, from a community where all farmers have formal school education, and where most of them have taken over their farms after the introduction of modern farming. These farmers had no memory of a pre-Green Revolution practices, such as seed selection and traditional seed management. In recent years, however, some of them have switched to organic farming because of declining profit margins in the high-input system. This has created a need for a different type of seed and also an organization to recover and reintroduce the lost traditional seeds as well as to re-establish a local seed system.

These two cases, one isolated from, and the other influenced by, the modern system required different organizational approaches to the seed management problems. In both cases, however, local seeds and on-farm selection were established as the platform on which to build. In Ethiopia, the felt problem was poverty and recurrent famines. Community leaders saw poor crop performance associated with poor seeds as one of the causes, but also knew about individuals who had good seeds and had a reputation of being excellent seed selectors. They decided to extend the good seeds through a credit scheme. Community seed banks were established and the local experts on traditional seed selection were used to identify good seeds for lending. Unlike genebanks, they were not concerned with conservation, but rather with the circulation of seeds. Like a commercial bank, they put their capital to work. They also had social concerns and gave priority to loan applicants who were poor and had a particular need for access to good seeds (and who needed protection against private moneylenders). The seed banks were owned by the community and controlled by democratically-elected community assemblies.

The first of these seed banks was established in 1988 and, within a few years, was replicated in most local districts of a region of close to four million people. The growth of the community seed banks continued after peace, in 1991. But now the challenge is twofold: the seed banks must develop in order to remain relevant; and they must protect their integrity and independence when government institutions and seed companies start appearing in the area.

During the war, people had no choice. They had to depend on their own resources. And they proved to themselves that the necessary skills and the required good seeds existed within their own community and could be used to solve their immediate problems. Currently, these Ethiopians do not seem to consider their seed effort as an emergency measure that can be phased out when government services and seed companies begin to function and, perhaps, to take over. They want to keep their seed banks as permanent institutions. Their challenge now is to realize the development potential of these institutions and the evolutionary potential of their seeds within the context of an opened-up economy. But the seed banks and their associated seed supply system are also a challenge for the scientific plant breeding system which is now being re-established in the area. That challenge need not render the local seed supply system obsolete by supplying better commercial seeds; scientists could opt to work with local farmer-breeders in order to ensure that seeds offered through the seed banks remain competitive.

The case from the Philippines, the Community Based Native Seed Research Centre (CONSERVE) in Mindanao, arose as a response to critical economic problems of the high-input system of rice cultivation. With increasing prices for inputs and decreasing prices for the produce, profit margins were shrinking, and farmers became dangerously dependent on moneylenders. Some individuals saw a switch to organic farming as the only way out. In that situation, farmers needed a new organization. Unlike the Ethiopian case, where community assemblies representing the entire community were the organizers and owners of the seed activities, this Philippine group was a minority, and membership was individual.

Starting in 1992, this group organized a search for local traditional seeds, which were recovered from farms in isolated remote areas. These seeds were multiplied and distributed to the members for on-farm evaluation and screening. After only two years of operation, I visited the project and found farmers discussing seeds with excitement. From an initial challenge of sorting and selecting among a great number of landraces offered to them, some had already started selecting within landraces, and some had started crossing varieties and keeping written records of what they were doing. They involved their wives and children. More than a change of seeds had occurred; there was a change of mind also. Before, these farmers grew modern varieties. Such varieties are supposed to be pure, and off-types are considered as impurities.

In case they were saving seeds, they had been taught to rogue the off-types before harvest to maintain varietal integrity. The change of mind involved seeing diversity in the field as a resource, rather than an impurity, and seeing themselves as active selectors, rather than passive receivers of ready-made varieties.

It is a mistake to consider this as a rejection of science. It is a withdrawal from the commercial seed system. If it is true that scientific plant breeders are working for farmers and not for seed companies, they might find organized farmer groups another outlet for their scientific achievements. That would require the development of participatory plant breeding methods.

Participatory Plant Breeding

Certain trends have made the world ripe for adoption of participatory plant breeding methods.

- The shielding of crops from environmental stress in high-input systems is facing increasing economic and ecological problems. Scientists are changing their attitudes, and a new paradigm is being formulated. Instead of modifying the environment to suit the requirement of high yielding varieties, the varieties need to be modified to suit the environment.
- The claim that modern varieties can be made broadly adapted and be superior across most farming environments within an ecogeographic region is being challenged. A number of recently published reports find varieties with specific local adaptation

to perform better when varieties are exposed to local stress environments.

- If relevant diversity exists in a locale, the combined action of natural and artificial selection within a local environment may be an efficient breeding method. Experiments show that this may work also in a fertilizer-intensive system.
- In recent years farmer groups working with local seeds have been organized all over the world. They are not primarily conservers of old seeds. They want their seeds to be improved, in an evolutionary, slow and steady way, and under their own control.
- Finally, participatory methods have been developed in order to facilitate the involvement of farmers together with scientists as active and equal partners in research to generate relevant farm technology. Such methods can be applied also to plant breeding.

To my knowledge, farmer attitudes to germplasm never include prejudice against modern varieties or any other form of exotic seeds. Commercial seeds often diffuse into areas where the traditional seed supply system is still predominant. Farmers try them with an open mind and adopt or reject them according to their own criteria. If grown and multiplied in the villages, diversity will start to appear within them and local reselection will be possible. In that way commercial varieties eventually might become like landraces. It is also commonly observed that farmers change and exchange seeds. A traditional farming system rarely functions as an environment for the static preservation of old landraces.

We are never come across a community where seed management is uniform. Often practices range from neglect to very simple mass selection, but with a few scattered individuals who devote an exceptional amount of effort to the maintenance or improvement of seed quality. These exceptional persons, very often women, may be the source of good seeds for others in the community.

Once such a community is organized for seed management and improvement, it becomes possible for it to establish links to scientific institutions. According to what I have gathered in a number of community visits, there is nothing in traditional attitudes that would make people reluctant to join a participatory breeding scheme — if it provides them with a wider selection base and prospects of progress through on farm evolutionary breeding.

Community resources for participatory approaches to seed management and breeding are not limited to indigenous culture and traditional practices. The educational status and experiences of modern farmers may also be turned into a resource for community action.

In the Philippine group, a number of the members were high school graduates and a few had a university degree. And moreover, most of them had a couple of decades' experience with modern input-intensive farming. Seed activities opened their minds towards the traditional societies, towards themselves and towards the modern world. The traditional societies supplied them with their seeds, and through the seeds, they learned to appreciate the values and achievements of these societies. They discovered their own potential, and saw that the outside world could bring more than technology packets: it could bring knowledge and ideas to be exploited and further developed by themselves.

In Conclusion: Using Biodiversity

The genebank collections serve as sources of genes for resistance to diseases and pests and little else. This limited use of germplasm collections may in the long run pose a serious threat the genetic heritage kept in genebanks. Who will keep paying for the maintenance of enormous numbers of seed samples that are hardly requested by anybody? Localized evolutionary breeding, however, will need the landraces with their specific adaptation and could, potentially, increase the demand on the genebanks tremendously.

Genebanks are organized to serve scientific plant breeding. In recent years genebanks are also being used to supply seeds for re-establishment of landraces that have been lost from disaster areas. This is done or planned for areas in Cambodia, Eritrea, Ethiopia, Somalia and Rwanda. But otherwise, local communities are not yet established as *bona fide* users of genebank materials. If farmers are organized and linked up to scientific institutions, it would be possible to establish a channel for return of relevant germplasm from genebanks to farm communities.

To some degree the direct return of landraces to areas from where they were originally collected and to other areas with similar agroenvironmental conditions may be warranted. However, it might be more useful if genebank materials are made available in the form of enhanced and enriched populations. Scientific institutions can synthesize such populations, but adaptation and selection should take place in farmers' fields.

Participatory Breeding and Selection

Alternative approaches for identifying cultivars that are acceptable to farmers have been suggested and tried by a number of authors. Chambers (1989) reviewed the small amount of work published at that time on providing farmers with varied genetic material. Published examples now encompass India, Rwanda, and Namibia in rice, beans and pearl millet.

In rice, Maurya *et al.* (1988) tested advanced lines with villagers in Uttar Pradesh and successfully identified superior material that was preferred by farmers. Joshi and Witcombe (1995) used farmer participatory methods to identify released rice cultivars that were not recommended in the research area. In Rwanda, farmers selected 21 varieties from a wide range of bean cultivars grown in their fields that were first selected by them in on-station trials. In Namibia, Lechner used farmer evaluation of pearl millet in on-station trials.

The farmers selected a cultivar that was subsequently released and became popular. In collaborative research between The International Crops Research Institute for the Semi-Arid Tropics (ICRISAT) and Rajasthan Agricultural University farmer participatory research was used to identify pearl millet cultivars suitable for Rajasthan (Weltzien *et al.* 1995, this volume).

All of these examples can best be defined as participatory varietal selection, since farmers were given near finished or finished products to test in their fields. In contrast, participatory plant breeding involves farmers selecting genotypes from segregating generations. There are few examples in the literature of participatory plant breeding. Sthapit *et al.* (this volume) have carried out PPB with farmers in Nepal to select chilling tolerant rice from F5 bulk families. Joshi and Witcombe have created a broadly based maize composite for participatory plant breeding in India, and the first selection by farmers will be in Gujarat in the *kharif* 1995.

Participatory Varietal Selection

Participatory varietal selection always has three phases:

- a means of *identifying farmers needs* in a cultivar,
- a *search* for suitable material to test with farmers, and
- *experimentation* on farmers fields.

Identification of Farmers Needs

A number of methods can be used, separately or in combination, to identify farmers needs. Important methods are:

- participatory rural appraisal (PRA),
- the examination of the type of crops in farmers fields at or near to maturity, or
- the pre-selection by farmers of cultivars by the inspection of trials of many entries grown on a research station or in farmers fields.

Employment of such methods will help to reduce the possibility that farmers will be given obviously unacceptable varieties to test.

Search Process

After the farmers' needs have been identified, the search process is carried out to identify suitable cultivars for testing with farmers.

Amongst already released cultivars: One method, employed by the authors in India, is to include in the search cultivars that have already been released. A key assumption made in participatory varietal selection on released cultivars is that cultivar replacement rates are lower than optimal because farmers have not been exposed to a range of new cultivars. It is therefore assumed that amongst the released cultivars there are ones that will be preferred by farmers over those they are currently growing. All that is required is to expose the farmers to the suitable cultivars for the project area that already exist, but have not been released or are not available in that area. For many crops in India, cultivars can be introduced from other states for a participatory varietal selection program since many cultivars have only been released in single states.

There is a considerable body of evidence to support the assumption that farmers are not rapidly adopting new cultivars because most cultivars under cultivation are old. There is also good evidence that only a few of the released cultivars are widely grown. For example, in wheat in India the average age of cultivars under breeder seed indent is 9 years, and the average of cultivars in certified seed production is 13 years in the three states of the KRIBP project, Gujarat, Madhya Pradesh and Rajasthan. The two most popular cultivars are Sonalika, and HD-2285 and these account for a large proportion of the area. However, for wheat, there is a good choice of cultivars as there have been 44 releases in the period 1984 to 1992 inclusive. In most crops, cultivars are on average older than those in wheat. For example, the average age of cultivars under breeder seed indent is 11 years in rice, 13 years in chickpea, 15 years in groundnut, 16 years in sorghum, and 17 years in maize.

Choosing from amongst released cultivars has the advantage that any non-governmental organization or governmental organization (NGO or GO, respectively) can, in principle, readily procure seeds in sufficient quantities for testing with farmers. If they are identified as being farmer-acceptable it should be much easier, than is the case for pre-release or breeders lines, to provide large quantities of seed to the farmers with little delay. Nonetheless, to increase the size of the basket of choices and exploit recent outputs from plant breeding research, pre-release cultivars might also be included in the search process and a number have identified as being suitable for testing with farmers in the KRIBP project. Some of the pre-release cultivars would be defined by others as advanced material.

Amongst advanced material: Maurya *et al.* 1988, after a PRA on farmers' needs, searched among characterized advanced lines to find suitable material for testing with farmers. In other cases, breeders have searched amongst pre-released cultivars (entries in advanced stages of testing) having local adaptation and have deliberately chosen material that represents a wide range of phenotypes. Sperling *et al.* (1993) and Lechner (pers. commun.) have successfully used farmer visits to research stations trials to identify suitable cultivars amongst the trial entries of non-released material.

Experimentation for PVS

Once genotypes or released cultivars have been identified and seed of them procured, various testing and evaluation systems can be employed that can vary greatly in terms of the extent of farmer participation. Many 'on farm trials are conducted almost entirely by researchers on farmers fields, so there is little or no involvement of farmers. At the other extreme, very limited inputs can be provided by outsiders such as scientists and development workers. To do so, farmers can be given a range of cultivars to grow for testing without intervention from outsiders. Outsider inputs in evaluating the material are also minimized by asking farmers in informal discussions which of the cultivars they like the most. Even these informal discussions can be avoided by merely waiting for demand for seed from farmers. On the basis of such discussions or demand from farmers, an NGO, a seed company, or a GO can make decisions on what seeds to provide to farmers. This informal research, with minimal outsider inputs, can be highly cost effective, and is recommended for NGOs with limited resources that have as a development objective the provision of seed of farmer-acceptable improved cultivars.

There are other methods that lie between the extremes of maximum and minimum inputs from outsiders. Scientists participation increases when farmers are asked to grow more than one introduced cultivar, since it necessitates an experimental design in which farmers, if unaided, can easily make planting errors. The scientists' contribution will also vary according to the quantity and quality of the data that is collected, and is greatest when quantitative estimates of yield using field-sampling techniques are employed. Therefore, a method that requires considerable input from scientists is when farmers are asked, with researcher help, to grow a set of cultivars, and quantitative data on their yield is taken by the researchers. However, most researchers when using participatory methods have asked farmers to grow only one introduction side by side with their local with no change in management, and have collected data that pertains to farmers perceptions of the cultivars.

The methods used by the authors are described in detail as an example of PVS. The varietal trials were carried out by farmers in Farmer Managed Participatory Research (FAMPAR) trials. The trials were divided into introductory and adaptive trials. The main difference between these two stages is that small quantities of seed were given to farmers in the introductory trials, but, to avoid overestimating the acceptability of cultivars to farmers, seed was sold at commercial rates in the adaptive trials. The trials were made as simple as possible. Each participating farmer was randomly assigned a single variety, and asked to grow it alongside the local variety in the same field. Each farmer was given two bags of seed, so that the plot could be resown, if required, from the second bag of seed. Farmers were asked to mark the plots and to do this they sometimes grew a row of different crop between the two cultivars. They were also asked not to change the management of the crop in any way. Enough seed was provided for an average plot size 100 m^2 which is much larger than that used in advanced on-station trials.

In the introductory trials, data were collected by means of Focus Group Discussions (FGDs) before and after harvest, on all aspects of the crop including taste, market value, threshing characteristics and storability. Evaluation was facilitated by 'farm walks in which the participating farmers visited each other's plots. All the cultivars could then be compared in the discussions, and it permitted the assessment of the reactions to each cultivar of all of the farmers that participated in the farm walks. The focused group discussions were followed by questionnaires completed for individual households, called household level questionnaires (HLQ), in which the household members reactions to the variety were assessed by means of a detailed questionnaire that included questions on post harvest traits, such as cooking quality and market value of the grain.

Results Obtained From Participatory Selection in KRIBP

Summary: Using these techniques, we have identified in the KRIBP project three cultivars of chickpea, two cultivars of rice, one cultivar of maize, and two cultivars of black gram that are markedly preferred by farmers. This has been achieved in only three years. One of the most revealing results is that recommended cultivars are rarely, if ever, preferred because their true recommendation domain is for areas where farmers grow crops in highly fertile soils where water is not limiting. Instead, all of the preferred cultivars, apart from one cultivar of rice and a national release in black gram, are introductions from outside of the three states in which the project area is situated.

This indicates that the recommendation domains of released cultivars have been defined too restrictively. Unfortunately, there is no mechanism of ensuring that once a cultivar is popular in one state of India that it is extensively tested in other states. An argument commonly used against the need to do this is that the material has already been tested in coordinated project trials. However, every case needs to be examined in detail. Often the number of locations in which an entry has been tested in any particular state will have been very small, and sometimes the trials that included the entry in question were rejected because the trial had excessive experimental error or because it failed.

PVS in rice: In 1993, introductory trials of rice were planned with 25 participating farmers in six villages, making a total of 150 farmers. Successful trials were conducted by 128 farmers, because some farmers failed to plant the seed.

In each village, five cultivars were grown and every cultivar was replicated across three to five farmers. The cultivars were Kalinga III, Sathi-34-36, Jaldi Dhan-1, Jaldi Dhan-3, and GR-3.

The farmers perceptions of Kalinga III, the most preferred variety, were compared in six villages. For yield, there was perfect agreement that Kalinga III was higher yielding in all villages in Madhya Pradesh and Gujarat. From observations of farmers fields, Kalinga III was seen to be considerably higher yielding. However, perceptions that Kalinga III was higher yielding than the local were far less marked in Rajasthan, but were always considered to be so by 50% or more of the farmers.

The probable reason is that the land is less sloping and more fertile in these Rajasthan villages, so there is a reduction in the advantage of Kalinga III, a cultivar that is highly adapted to low input conditions.

PVS in chickpea: In *rabi* 1992/1993, five chickpea cultivars were grown in six villages. In each village, each cultivar was grown by four farmers to give a total of 120 farmers. After the harvest in 1993, a HLQ was conducted and farmers were asked if they would grow their chickpea cultivar again. Three cultivars, ICCV 2, ICCV 88202 and ICCV 10, were selected by a good proportion of the farmers, but ICCC 4 the official released variety was not liked. Differences between the cultivars were less clear when the cultivar choice was restricted to the cultivar the farmer had grown, than when choice was restricted to the remaining cultivars. Probably the availability of seed of the cultivar the farmer has grown greatly influences the decision in favour of regrowing it.

The second-year adoption rates of the preferred cultivars were found for the cultivars by interviewing farmers in two villages, (those that had participated in trials in 1993/94), at the end of the 1994/95 season. The differences in adoption rates into the second year showed that ICCC 4 was even less liked than was indicated by the HLQ. ICCV 37 was resown by two of the four farmers and no seed was given to others. Adoption rates were higher for the three preferred cultivars in the HLQ. ICCV 88202 and ICCV 2 were more preferred than ICCV 10 in terms of both adoption rate and the number of recipients of seed and this result agreed with those from the HLQ.

Since the two surveys had only one village in common, this agreement was even more impressive. The three farmer-preferred cultivars will spread with farmer-to-farmer seed supply, and the area under these cultivars will increase quite rapidly, even if no further seed is supplied from outside. Multiplication rates are conservatively estimated at between two and three times and there is a high rate of spread of seed to new areas.

The results from the three different surveys all agree in finding ICCV 2, ICCV 88202 and ICCV 10 as the preferred varieties, but the order of preference changed somewhat. This is perhaps unsurprising as the three cultivars all have markedly different characteristics. ICCV 2 is very early kabuli type, the seeds of which fetch a higher market price. ICCV 88202 is a very early desi type, and ICCV 10 is later than ICCV 88202 but higher yielding. Different villages may have different preferences according to such factors as access to markets and soil types. By exposing the farmers to a diverse range of genotypes a number of cultivars have been adopted, and biodiversity has been increased since they are all being adopted and partially replacing the uniform single 'landrace cultivar, Dahod yellow, that was exclusively cultivated in the area before these introductions.

Participatory Plant Breeding

Participatory varietal selection has been extended to participatory plant breeding (PPB) on the assumption that if it is desirable to involve farmers in selection of cultivars then why wait until there are finished products? Farmers can be involved at a much earlier stage whilst material is still segregating. However, participatory plant breeding is more resource consuming than PVS, and hence the first recourse should be to the least expensive method. PPB has to be used when PVS has been tried and failed, or when the search process has failed to identify any suitable candidate cultivars.

The methods used in participatory plant breeding are poorly documented since there are no reports in the literature of a completed participatory plant breeding program. Sthapit *et al.* (this volume) have used F5 bulk families as the starting point for their participatory breeding program. These were derived from seed harvested from F4 families that were grown in the farmers village. The breeding scheme is at the F5 stage in the monsoon season of 1995, and it is intended to monitor progress in the farmers fields until a finished product is produced. In contrast, Thakur (1995) has screened material in farmers' fields at the F2 stage, but subsequent generations have been grown by researchers. The authors, in collaboration with Dr. Goyal of Gujarat Agricultural University, are starting a participatory plant breeding in maize with the fourth random mating generation of a composite created from six farmer-acceptable open-pollinated cultivars. In Ethiopia, farmer enhancement of landraces by mass selection has been done in collaboration with scientists from the Plant Genetic Resources Centre. In participatory plant breeding in rice in the Philippines, farmers are involved in selecting from progeny of crosses between traditional and improved cultivars but, unfortunately, the methods used are not described in detail.

A range of participatory plant breeding methods are possible with predominantly self-pollinating crops, and they have been ordered by degree of farmer participation. The methods vary according to which generations are grown by farmers and by the extent of researcher participation. The method with the greatest farmer participation and the greatest number of generations requires little breeder input during the selection stages.

However, an essential role is played by breeders and participatory plant breeding is not intended to make plant breeders redundant. In all of the methods, the plant breeder is the facilitator of the research. Only the plant breeder can make the crosses between the parents and have the essential understanding of the underlying genetics in the segregating generations. Moreover, only the plant breeder has the knowledge of the official release system, and cultivar release is still a very desirable end product to make the results of the participatory research more widely available.

For predominantly open-pollinating crops, plant breeders can create composites in isolation and give the third or fourth random mating to farmers for mass selection. Large plots of composite have to be grown by farmers, or small plots need to be isolated by time of flowering or distance from other plots of the same crop. Because of these constraints, it is difficult to carry out the breeding scheme in many locations. Mass selection can be done with or without an off-season generation controlled by the plant breeder. However, to breed for wider adaptation, plant breeders can recombine selections from different farmers in the off-season. Although it is likely to reduce the progress made by selection over the best farmers selection, it avoids the risk of continuing a population from a poor selection by one farmer or from a population where outcrossing with other farmers crops happened to be higher than expected.

Greater plant breeder input is possible by using a method of progeny testing. Plant breeders can produce progeny before giving material to farmers, and can produce progeny between generations of farmer selection. In the most extremely breeder-oriented system, the farmers grow progeny trials of full-sib or S1 families, and the breeder recombines from remnant seed the farmer-preferred progeny.

Impact of Farmer Participatory Research on Biodiversity

Biodiversity in Crops

Biodiversity in crops is very difficult to define and a number of simplifications are assumed in the following discussions. The degree of genetic relatedness of one cultivar to another is not considered. It is assumed that if one cultivar partially replaces another, or several cultivars replace one cultivar, there is an increase in biodiversity. However, the degree of increase will vary considerably depending on the genetic dissimilarities among a range of cultivars that are adopted, or the genetic dissimilarity between an existing cultivar and one that partially replaces it.

Moreover, biodiversity is not only a function of the total number of cultivars. Given two agroecosystems with an equal number of cultivars, the agroecosystem having a large proportion of the area occupied by a single cultivar is more genetically vulnerable than one where the cultivars occupy nearly equal areas. Great difficulties then arise between balancing the total number of cultivars against how equally they occupy the cultivated area. For example, in the case of rice in KRIBP, the question can be asked: does the adoption of Kalinga III reduce or increase biodiversity? Assuming it does not replace any single landrace completely, has it contributed to an increase in biodiversity by increasing the number of cultivars grown, or has it reduced biodiversity by occupying a large proportion of the area where previously several landraces were grown?

Biodiversity can be over both space and time. When one cultivar totally replaces another, there is an increase in biodiversity over time. There is also a temporary increase in biodiversity over space until the replacement is complete because, while replacement is occurring, there are two cultivars in farmers fields instead of one. The pattern by which this replacement takes place, from many or only a few foci, will also be important. We can assume that there is greater biodiversity when the new cultivars spreads from many foci. An example of this type of spread is seen in the case for new chickpea cultivars in KRIBP. The spread from many foci will give a more complex pattern between farmers fields, providing a useful increase in biodiversity. The vulnerability of a crop to a disease is reduced when there are many field-to-field differences than when there are few. This strategy of field-to-field variability has been recommended by Priestly and Bayles (1980, 1982). However, regional variation in cultivar diversity is also suggested as useful in disease control by Frey *et al.* (1977).

When participatory research increases replacement rates, and thus reduces the longevity of individual cultivars, biodiversity is increased over time. It is again a useful increase in biodiversity since pathogens and pests are exposed to a particular genotype for less time and have less chance to overcome host plant resistances.

Participatory Varietal Selection

When farmers are exposed to the 'basket of choices of a range of new cultivars in a participatory selection program, the outcome in a specific region may be an increase or a reduction in biodiversity. The situation is complex and changes in biodiversity depend on existing variability in farmers fields, the variability in the new cultivars offered to farmers and their acceptability, and the variability in the target environment, both physical and socio-economic.

The greater the physical diversity in the environment, the more likely it is that more than one cultivar will be adopted by farmers. Diversity of economic use will also make it more likely that several cultivars are found acceptable and are adopted. Often farmers will prefer different grain types for different purposes. High-yielding cultivars with poor quality grain may be grown as a cash crop or to reduce risk, whilst cultivars with high quality grain will be grown for home consumption, and for special social and religious occasions.

Nonetheless, when conditions are right for several or many cultivars to be adopted then it is likely that great diversity already exists in farmers fields. When PVS is employed in areas where highly variable landraces are grown and there is no or little adoption of improved cultivars then its success will reduce biodiversity. This dilemma is faced by NGOs that wish to conserve biodiversity and help resource-poor farmers. One example of the problem is given by Cromwell and Wiggins (1993). An NGO, the Save the Children Federation in The Gambia, were faced with *"the dilemma of seeing just one of its introduced rice varieties almost completely replace the range of local varieties previously grown, which were no longer suitable because of declining rainfall."*

Biodiversity can also be considered over a wider area such as at the national level. We can assume that the widespread adoption of participatory methods will increase the replacement rate of cultivars, so that the average age of cultivars grown by farmers will be reduced and biodiversity over time increased.

It is also likely that adoption ceilings of improved cultivars will increase at the cost of reduced biodiversity. Overcoming inefficiencies that limit the adoption of improved cultivars to relatively small areas will be balanced by the adoption in any area of a greater number of cultivars. PVS reduces biodiversity when a cultivar is adopted over a wider area, and a good example is the rapid adoption of Kalinga III in western India following PVS by farmers when the cultivar was previously only released in eastern India.

However, in the longer term, PVS should have only beneficial effects on biodiversity. If many more farmers are exposed to many more cultivars, the number of cultivars adopted will increase and the patchwork of cultivars between fields, districts and regions will increase in complexity.

Participatory Plant Breeding

The impact on biodiversity of participatory plant breeding, in contrast to PVS, is easy to predict since PPB will increase biodiversity under nearly all circumstance. When compared to PVS, the increase in biodiversity will be at both the intra-and inter-varietal level. The effects of PPB will be more uneven than with PVS, with a very high impact on biodiversity in the participating villages, and an impact that elsewhere is restricted to cultivars that spread from village to village.

In predominantly self-pollinating crops, the adoption of PPB brings a change of methodology. Instead of conventional pedigree breeding, bulk population breeding is employed whereby farmers mass select within segregating populations, such as the F4 bulk families used by Sthapit *et al.* in Nepal. Hence PPB increases biodiversity at the intra-cultivar level overcoming the 'disease of cultivar uniformity. Intra-cultivar variability in the form of multiline cultivars is recommended as a strategy for reducing disease (Browning and Frey, 1981), but in this method intra-cultivar diversity is deliberately minimized apart from variability for disease resistance genes. The intra-cultivar diversity generated from PPB is more akin to a varietal mixture and such mixtures have been effective in disease control.

More difficult to monitor spread of any variety in participating villages because of biodiversity created. Concern on part of breeders that farmers may have produced many varieties so which one to identify and promote? However, any improved preferred variety is better than none.

PPB is also a logical second stage to PVS, and if the appropriate breeding methods are employed then it comes closest to the ideals of genetic conservation. After PVS has been successful, the farmer-preferred cultivars can be crossed to other materials for farmers to select in the progeny. Breeding strategies will involve crossing the cultivar identified by participatory varietal selection (termed the PVS cultivar) with landraces and with high-yielding released cultivars. In the first strategy, the landrace is chosen as a parent to give genes for adaptation, and, in the second, a released cultivar is chosen to give genes for high yield potential. When landrace x PVS cultivar crosses are used and there is maximal farmer input in the breeding. then we have a breeding strategy that most closely resembles *in situ* conservation of landraces. Farmer experimentation on naturally existing genetic variation has produced landraces, and this method enhances such farmer experimentation. Genetic variation is increased by the hybridization between the landraces and the PVS cultivar, and selection procedures by farmers and farmer awareness are enhanced by interaction with scientists. Nonetheless, in this process there is a possibility that some useful genes present in the landraces will be lost so *ex situ* conservation will still be desirable. Certainly, if there is a desire to wholly preserve the existing landraces, then *ex situ* conservation is essential. We can say that *in situ* PPB conserves genetic resources in farmers fields whereas *ex situ* conservation preserves genetic resources.

Seed Selection and Plant Breeding

Educational Importance of Seed Selection.—Perhaps nowhere else can the children be interested more easily than in the selecting of superior garden vegetables. In France, we are told, they have the finest vegetables in the world, and this is so because they practise the most careful seed selection. And strange to say, in France the children are taught to select and prepare the seeds for the garden. With us it is very different. The American gardener sells the superior vegetables, especially the earlier ones, and then he sells to the seed houses the seed from those that were not fit for market. A strange " survival of the fittest" we practise ! This subject of better fruit and better garden vegetables for the farmer's own table has a far-reaching sociological meaning. We shall never make farm life what it should be until we enable the mothers on the farms to have a better time than do mothers in towns and cities. This means that the women on the farms must live in houses with modern conveniences and have plenty of first-class things to cook. More fruit and better garden vegetables would stop the longing of many a woman to go to town to live. Enable the mother on the farm to become conscious that she is free from her town sister's nagging worry as to where the next meal is coming from; make her conscious that she lives in a good house; that she has more vegetables, fruit, flowers, sunshine and birds; make her conscious that the growing of vegetables has a far-reaching educational significance and you will increase her desire to " stay on the old farm a while longer." A woman in the country who has these advantages and who is conscious of them will not " be in a hurry to go."

Teachers should remember that both the, psychologists and the sociologists place great emphasis on the importance of the development of forethought. Nothing found in the present school course equals seed selection for the development of forethought. The power of forethought has been said to be the faculty that most clearly differentiates man from the lower animals and civilized man from the lower races.

Seed Selection for Fall or Spring.—Seed selection makes an equally good subject for those who begin the work in September and for those who begin to teach agriculture in the spring, though the hot-bed, the hot-house, and the school garden are the logically and seasonably correct subjects for' the work in the spring. If seed selection is begun in the spring there may be discussions of where the best seeds are to be obtained and of the work of our seedsmen who up to this time have been our principal plant breeders. Having secured the best seed to be obtained, and having planted it on good ground and by the most approved methods, the teacher may next lead the children to watch the school or home garden. As soon as the first vegetables appear the best should be marked and saved to bear seed. The marking is done by setting a stake on either side of the best appearing radish, lettuce, or whatever the plant may be. This vegetable may be watched through the season, though if it be lettuce, radish, or some other vegetable which matures its seed the first season, all that is necessary to do is to cultivate it with the others, and let it stand to mature seed. But if it be cabbage, beet, carrot, parsnip, turnip or some other plant which does not go to seed the first season, then it must be pulled in the fall, packed in damp sand and kept over winter in a cool place. The next season the root (if cabbage stump or stalk) is to be replanted and let go to seed. The cabbage head may be used and only the stem and roots wintered over.

While doing this work many questions will arise. How to plant, how to cultivate, how best to select and preserve seed, how to fertilize, what is the best kind of soil, and others will be asked. These are good; they come from the instinct of curiosity. They are of most educational value if answers can be put off until the proper time and each answered as answers are needed.

Propagation by Roots and Cutting.—During the fall months there is much to do on the farms, with plants that are propagated from roots and cuttings. Cuttings of grape vines with four buds on each may be placed in the moist ground with the butt or large ends up. It is best to have them slant about forty-five degrees to the earth surface. They should be covered with moist soil and left until spring, when they are to be dug up and stuck in the ground, large end downward, and far enough to have the soil cover the two lower buds. This gives two buds for roots, and two for leaves. Young trees may be taken up in the fall and heeled in for the winter or for early spring planting. Others may be taken up and heeled in moist sand in a box to be stored in some convenient place where it may be gotten at in February or March. It is during the slack times of these months that the grafting is done. But before I tell how to graft, I need to tell how to start an apple tree. We must gather the seeds from the apple and plant them in a window box, a hot-bed or hot-house before they become thoroughly dry. We may put them in moist sand and set the box out where the seeds will keep cold until spring and where they will freeze so that the little plant may the easier get out of the hard shell.

However, many of the apple seeds will sprout at once if placed in good soil kept well dampened in the window box. One teacher used a chalk box for starting her apple seeds. The apple seed has come from cross-pollination and hence we are not sure just what parents the little tree has, that is, we do not know what kind of apples it will bear. Now to insure its bearing the kind of apples that we want, we graft a limb of some known variety into the stump of the little tree. For this work in grafting it is necessary to know something of the cambium layer. This is the part of the dicotyledonous plant producing growth. We call it parenchyma (paent) tissue. It is from the cambium layer that the fibrovascular bundles develop. These fibrovascular bundles are the sap pipes for the plant. They carry the sap up and down to nourish the different parts or to be mixed with carbon and more thoroughly digested in the leaves. This cambium layer is the white layer just under the bark. It is the part that when watery enough enables us to slip the bark off to make a willow whistle.

When grafting, we must be sure that on one side at least, the cambium of the stock and the scion (branch) just exactly meet. Then we put on grafting wax to keep out bacteria and the spores of fungous diseases. We cover the wax with cloth or corn-husks to keep it in place until our graft is " set."

If we grow a peach tree, we plant the seed as we did the apple, but instead of grafting, we bud the peach. This is easier for all that we have to do is to cut off a leaf bud and slip it under the bark of the tree. Of course we must make cambium meet cambium as in the case of the apple graft. We must bind up the wound in some way so as to keep our bacteria and the spores of diseases. After our bud is set and begins to grow, we cut off the little tree above the bud unless we wish it to bear more than one kind of peaches. Then when the proper time comes we must move our little trees from their nurseries to the places where we wish them to stand when grown.

This process of moving we call transplanting. Most farmers buy their nursery stock and hence transplanting is for most farmers more important than budding or grafting. It is becoming more and more common to find one of the redirected schools doing the plant-breeding work. These schools are able to have little nurseries in connection with their school gardens and are able to furnish all of the trees that the farmers of the district need to replace trees in their orchards. There are:

Ten Rules for Transplanting

1. Make the holes large enough to receive the roots easily.
2. Make the holes deep enough so that the soil will come just above the swelling made by the grafting.
3. Unless you are transplanting forest trees, put a shovel full of well-rotted manure in the bottom of the hole.
4. Take up the little trees with the least possible harm to their roots.
5. Trim off with a sharp knife all mangled and broken roots.
6. Cut off enough of the branches to balance the injured root system.
7. Dip the roots in water before planting. The roots should be kept damp; this may be done by coating them with wet clay soil.
8. If the soil is dry, add water with the soil until the hole is about half full.
9. Pack the dirt solidly around the roots of the little tree. The soil does not need packing at the top.
10. If transplanting in sod or lawn, do not heap the dirt around the tree but instead leave a basin to catch water and avoid danger when mowing.

Seed Envelopes and Cards

During September and October, whether the seed work was begun in the spring or in the fall, the children should be induced to gather the seeds, thoroughly dry them, and then put them away in paper sacks, where mice and insects cannot get at them. Later the children of some of the grades may prepare envelopes for the seeds. The envelopes may be obtained from seed houses, or ordinary correspondence envelopes may be used. On these envelopes the children are to write the name of the plant from which the seed came, variety if known, whether early or late, time of planting, best kind of soil, and as much of how to plant and cultivate as space will permit. These envelopes may be pinned or stapled on to large pieces of stiff pasteboard and used for part of the school exhibits.

Booklets Correlate with Reading, Writing and Spelling

The pupils should start booklets on plants, plant breeding, seed selection or some special phase of the study. They may be most of the school year making the booklets, but whenever they get a page good enough to keep, they should have a place in which to keep it. The booklets may contain the, history of the plant, its value for food, how to breed it, how to cook and prepare it, and how to keep it. A list of the different varieties grown in that section of the country, with the relative advantages and disadvantages of each, makes splendid booklet material. The booklet work may be done as part of the English, writing, reading or spelling lessons. It should not be done as part of the work in agriculture to the exclusion of the work that deals with things out-of-doors. Covers may be made for the booklets as part of the work in drawing. The farm journals, the catalogues of the seed houses, and the nurseries furnish illustrations, pictures, letters for lettering, and valuable suggestions for subject matter. Sometimes it is necessary to use some of the government bulletins for pictures or illustrations; these do not need to be destroyed but copies of the illustrations may be made. This booklet work should make the child feel that there is much he should read. The booklet work should make him desire to write better, to spell more accurately, to acquire a larger and more accurate vocabulary and a better sentence structure. These are necessary to enable him to prepare a booklet of which he may be proud as he sees it in the school exhibit on commencement day, at the fall festival and fairs.

Where there is an old gardener who is an interesting talker it is often advisable for the teacher to get him to come to the school-house, or, better, take the pupils to his place, and have him give the children a talk on growing some particular vegetable, how to prepare the seed-bed, plant and cultivate, selection and care of seed, or any one of a hundred interesting things for children to learn. His talk, or talks, make valuable material for English lessons and then for a page or two of the pupil's booklet.

Some government bulletin or some good book may be brought from the library and one of the brighter or more interested pupils allowed to read what it has to say on the particular plant under discussion, and then this pupil may report to the class or the school, and here again we have material for composition lessons and perhaps another page or so for the booklets. The child may talk of these discussions at home, and from that some parent may want to see the bulletin or booklet and thus home and school are drawn closer together and a little child is leading them.

Correlate the Work with Geography and Botany

When the history of the plant is being looked up, for a page in the book-let, the teacher should not miss the opportunity to have the pupil apply some of his work in geography. If he has learned to use maps, he will easily be able to point out on the map the place where the plant originated and, if he is deficient in this ability to use maps, here is his opportunity to become somewhat more proficient. If the teacher has studied botany or is willing to look the subject up, she may do some valuable work in classification of roots, underground stems, tubers, and so on. Children should learn to call things by their right names. Then, too, the flowers, the parts of the flowers and their right names may be taught to children of the lower grades.

The teacher should use the names and teach the children to call the outer row of green leaves, or scales, bract or calyx, as the case may be. The coloured row should be called corolla if it is the corolla. The stamens and pistil and their parts—anther, filament, style and stigma—should all be called by their right names. The characteristics and names of six or ten of the leading families should be learned by pupils before they reach the high school. They should learn to name and recognize the representative members of these families. Farmers frequently listen to lectures from which they derive much more benefit if they have some knowledge of how plants are classified. Rotation of crops should re-quire a change of plant families. Corn, wheat and oats are members of the grass family (Graminece) and, while a succession of these three crops may enable the farmer to handle his labour and soil to better advantage, yet there is no benefit from a succession of the different members of the grass family to be compared with the benefit which a farmer gets if he follows members of the grass family by members of the pea, or Leguminosce family. The rose, the Unibelliferoe (carrot), the Cruciferoe, and the Compositoe, the nightshade or Solanaceoe families are families with which a farmer has much to do.

Species and Creationism

Then, too, life will mean more and when the boy becomes a man he may the better cooperate with the forces of nature if his school has put him into possession of the accumulated knowledge of the ages as to how his plants and animals came to be what they are. The first botanists had little idea of evolution and hence tried to classify plants on the theory that species of both plants and animals were created separately and were distinct from the beginning. Linnaeus voices the belief of his age and of former ages, when he gives us his now famous sentence, " We reckon so many species as there were distinct forms created in the beginning." In 1691, Camerarius discovered the fact that plants have sex, and in 1719; Thomas Fairchild, an English gardener, produced the first recorded hybrid. During these years and the following years, men worked out elaborate systems of classifications for the plants, on the theory of special creations. They reasoned that since plants were created distinct, therefore we may classify them into species the pollen of which will not fertilize or at least will not enable the plant to mature seed which will grow. Hence species theoretically stands for plants which we cannot hybridize. But as was to be expected, the early botanists made many mistakes, and hence the term " species " helps us little in determining what will and what will not hybridize.

In the beginning of the nineteenth century, Lamark advanced the theory that living things have descended from a common ancestor. He assumed that the influence of the environment was able to change the characteristics of plants and animals. His evidence was not sufficient to enable Lamark to convince the people of his day and hence the theory of special creations or Creationism, was very generally held until 1859, when Charles Darwin published his " Origin of Species."

Evolution and the Origin of Species,, Darwin's evidence for proving evolution was so abundant and his handling of it was so masterful that it has remained unshaken. His being the last " in the line " enabled him to receive more popular credit than he deserved, so far as the discovery of the theory of a common descent is concerned. Darwin proved that the origin of species in the past was the same as the origin of new varieties in the present. Of course men objected on the ground that we do not observe actual, specific changes of species. To meet this objection, Darwin advanced the theory that the changes are so slight that we fail at any given time to notice them. But they are accumulative, and hence in ages they become very notice-able. But to this physicists and astronomers objected on the ground that the time for the creation of our varied plant and animal life would be too long. Lord Kelvin and others by figuring on the radiation of heat from the earth, on the deposits of calcareous rocks, on the amount of salt in the sea, etc., were led to object that the time for the evolution of all species from a common ancestor by the slight changes assumed by Darwin was entirely too short. But though there were objections to Darwin's explanations, still his theory of a common descent had come to stay. To-day the leading scientists believe that. Then, too, Darwin did much to establish and explain variation. He found some variations for which he could not account and he called such " sports." He gave us the phrase " natural selection," to which Spencer added " survival of the fittest," both very useful intellectual tools for the husbandman. And Darwin made it necessary for us to reclassify our plants if not to discover a better explanation for the way one species comes from another. Darwin also gave us a new scientific age in which we demand " sufficient evidence." This means that we ask the scientist to give us a cause big enough to produce the result which he claims for it.

Darwin did much to bring that happy day when people in the country may live in the " certainty of science." Country people are more dependent upon nature than are town people and hence country people should know more of life and death, of lightning and wind, and of how things come about. Much of the unhappiness in the country comes from superstition. For superstition we must substitute the " certainty of science " or the demand for sufficient evidence. Then, too, the farmer is ever working with variations, natural selection and artificial selection, and he is trying to secure the survival of the fittest. But, thankful as the farmer should be to Darwin for the ideas and the phrases and evidence for common descent, survival of the fittest, natural selection, and the age that demands sufficient evidence, he has a right to demand of his school—especially of his high school—that it put him into possession of a better explanation for much of the phenomena which he observes around him.

Variation and the Constancy of Species

The leading scientists of today, especially the plant breeders, are at a loss to find illustrations of variations in the direction of evolution. They find that plants are changed by dropping something, such as part of their colour, or the width, length, or texture of an organ. But this is apt to be reversion instead of progression. The plant breeders who have been dealing with thousands upon thousands of plants have very generally come to believe that species may, like individuals, have their birth, lifetime and death, that the struggle is between species and hence plants become extinct by species and not by individuals. This belief is strengthened by the discoveries of the old beloved teacher of mathematics, Gregory Mendel.

Mendelism

Mendel was monk and abbot of the monastery at Briinn. Ile bred and crossed peas and carefully tabulated results, and from his work in his little garden we have some of the most helpful facts made known in modem times. Mendel found order in nature. He found certainty where others could see nothing but uncertainty and disorder. He found mathematical order in heredity. After eight years' work he published his observations in the Proceedings of the Natural History Society of Briinn. But his account of his experiments attracted very little attention until rediscovered in 1890 by Hugo DeVries and others.

Mendel gave us the laws of heredity. By use of Mendel's laws, we learn that we get from a plant just what we put into it. Some of the scientists are led to believe that there is no variation at all but just possibilities of different combinations of factors. To describe Mendelism, terms similar to those used by the chemist were necessary and hence the terms unit-characteristics and factors were coined. By unit-characters we mean certain parts of plants or animals which are inherited as units; for example, colour, form, size, horns, hairs, open or closed fibrovascular bundles, bark, shape of leaf, etc.

One of the corn club boys made very good use of his knowledge of unit-characters. He knew that others had difficulty in trying to get high protein corn to yield heavily. But he noticed that some ears have kernels which when in test give three to five rootlets of equal size while others give one large one and few if any smaller ones. From this he reasoned that by using his high protein corn and selecting for heavy rooters, he might get better yields. He won a trip to Washington over his competitors.

But the scientists soon discovered the fact that unit-characters are too complex. They resemble the compounds or molecules of the chemist. Breeders need a word that denominates the phenomena reduced to its simplest terms, and for that they adopted the name factors. Colour, for example, may be yellow, red, blue, green or a combination. Hence in plant breeding we may be dealing with the yellow or the red only, and instead of calling it a unit-character we call it a factor.

Heredity

What order did Mendel find? Others had noticed that the young resemble the parents and hence there is heredity. They had noticed that no two plants, parts of the plant or animal are alike and hence there is variation. Others had noticed that this is most certainly true where there are two parents, and not one as in plants which multiply by division such as the bacteria and some of the algae.

But Mendel discovered that the ratio of the young that resemble either parent is as three to one. He noticed that where some things come together, as black and white, the black is apt to hide the white, and hence he called the black dominant and the white recessive. If we cross two hybrid black and white seeds or animals, what do we get? If we try the experiment often enough, we are sure to average a three to one ratio, that is, we get: 1 black, plus 2 mixed, plus 1 white.

If we use one characteristic such as colour with two factors, such as white and black, we have four possibilities. If we use two characteristics such as colour and size with two factors each, such as black and white with large and small, we have the dihybrid (3 + 1)2 or sixteen possibilities. If we use three characteristics with two factors each we have the trihybrid (3 + 1)3 or sixty-four possibilities, and if we use ten characteristics we have (3 + 1)1° or 1,048,576 possibilities. These are sufficient to show that by using nothing but different combinations of the factors of the characteristics of the parents we are able to get a wonderfully varied plant and animal life. But we get them with the certainty of mathematics. " Whatsoever a man soweth that shall he also reap."

These discoveries, made by men who were trying to work out and explain the Mendelian experiments, help to confirm the belief that species are fixed, have their birth, life and death as naturally as individuals. Plant breeders were unable to get results by following the theory of the slow variations of Darwin.

Nilsson says of his work at Svalof in Sweden: " Rigorous selection pursued for five years had produced only a relative uniformity; we could not show a single new and constant variety-character. And most of all, it was evident that our selected varieties, left to themselves for a year or two, unquestionably fell back to the condition of a mixture of the original varieties.

> *"Evidently we were unable to produce what the Swedish farmers wanted—better varieties, which would be constant. It was obvious that we must find a new method."*

DeVries and the Mutation Theory

Not only among the plant breeders was it necessary to get a new method, but among the scientists it was necessary to have a better explanation of what plants do in heredity. This led Hugo DeVries of Holland to announce his now famous Mutation Theory. By this we are taught that while it is true that most plants produce young that obey the Mendelian ratios, yet at irregular intervals and for reasons as yet unexplained some plants give off variations which do not obey the Mendelian laws—variations or sports, as Darwin called them, which are not to be explained by a combination of the factors of the parents. Now if these sports be prepotentthat is, if they have the power to hand on to their young and their young's young their peculiar variation—we have what DeVries calls a mutant. This means a sudden creation of a new species. This does not contradict Darwin's common descent of species, but it offers a better explanation of how new species arise.

Mutation and Plant Breeding

The mutation theory gave a new impetus and a new method for plant breeding. With this theory a man is to deal with individual plants. He is to watch for new species or mutations and then, as in nature, there is to be a struggle among species to win his favour. It was the use of this new theory and the new method growing out of it that enabled Nilsson to become Europe's leading breeder of agricultural plants. Says Nilsson: " It is a fact well authenticated at Svalof that mutations appear from time to time in our cultivated plants. Furthermore, we have found that spontaneous fecundation is much commoner than we had supposed."

Some Examples of Mutants

Many illustrations could be given of what are believed to be mutants. The Concord grape came from a seedling of one of the wild grapes of New England, and from the Concord have come the Worden, Moor's Early, Pocklington, Rockland and others. Nectarines and apricots appear to have mutated from peaches and plums respectively. Other mutants are:

The Hubbard squash, the Morgan horse, DeVries' primroses, Swedish wheats (Extra Square-head and Sol), oats (Klock II and Seger), barley, peas (Concordia), and many American grains and fruits. But the reader needs to be cautioned that there is a group of scientists who do not believe in mutations and they claim that these are hybrids. We lack analysis of their parents and histories of their ancestors so that we cannot prove for a certainty that these are mutations. However, the theory of mutations sets us to watch for the exceptional individuals. This theory also has led the breeders to use single plants and their progeny for their breeding experiments, and from this use of individuals we have some very valuable grains and fruits.

It is my belief that mutants are appearing in the farmers' fields and yards much oftener than we are aware of. Mutants in plants and domestic animals are of value only as we are able to recognize them and perpetuate them. Left to themselves, mutants soon mix with others and are lost to us forever. We need a generation of young people put into possession of the scientists' contribution to civilization so that we may carry on the work where our parents leave off.

"It is no stretch of the imagination," says Dean Davenport, " nor is it a chimerical dream to say that the students of our better schools, aided by their teachers, can, if they will, do more to further improve many of our cultivated plants than can the farmers themselves. They have the time and can acquire the skill—things which are difficult to secure to the man who is busy in active commercial life."

Possibilities of Plant Breeding

It should be the aim of agriculture in the public schools to develop and inspire a practical plant breeder for every community. Wonderful things have been accomplished in plant breeding and more wonderful things seem just ahead of us. Napoleon had his men breed the sugar beet, and in a very few years they bred it from a watery garden vegetable with 3 per cent of sugar in it to the commercial beet containing 16 per cent of sugar. It is reported that the Experiment Station of South Dakota has a beet with 29 per cent of sugar.

Porto Rico exports $25,000,000 worth of sugar from cane averaging 11 per cent of sucrose. The United States Experiment Station in Porto Rico, by crossing the native cane with a British West India cane, succeeded in getting a cane that averages 21 per cent of sucrose. Man has bred the apple from the little, puckery, gnarled crab to the almost numberless shapes, colours and luscious fruits found in our orchards. He has bred the seed from the grape and the orange, and created a new fruit, called the grape-fruit.

Luther Burbank, whose biography every country boy should read, has proven that we may breed the thorns from the blackberry, the rose and the cactus. Mendel, the Austrian monk, proved that we can find in plant breeding mathematical order of which most cultured people never dreamed. Professor Hayes, of Minnesota, raised the protein content of barley, and Professor Hopkins, of the University of Illinois, raised and lowered the chemical contents of corn almost at will. Professor Thomas Hunt, in that most valuable book for the farmer boy to read, "The Cereals in America," quotes Burbank as follows: " The vast possibilities of plant breeding can hardly be estimated. It would not be difficult for one man to breed a new rye, wheat, barley, oats or rice which would produce one grain more to each head, or a corn which would produce an extra kernel to each ear, another potato on each plant, or an apple, plum, orange or nut to each tree. What would be the result? In five staples only in the United States alone the inexhaustible forces of nature would produce annually without effort and without cost::

5,200,000 extra bushels of corn.

15,000,000 extra bushels of wheat.

20,000,000 extra bushels of oats.

1,500,000 extra bushels of barley.

21,000,000 extra bushels of potatoes.

> *"But such vast possibilities are not alone for one year, or for one time or race, but are beneficent legacies for every man, woman, or child who shall ever inhabit the earth. And who can estimate the elevating and refining influence and moral value of flowers, with all of their graceful forms and bewitching shades and combinations for colour and exquisitely varied perfumes? These silent influences are unconsciously felt even by those who do not appreciate them consciously."*

Man may by care and cultivation make plants mature better seeds and fruits temporarily, but, by breeding, new plants may be created which will produce better always, in all places, and for all time to come. Heredity is a force that works without expense. The farmer who lacks the skill or time and talent for plant breeding can afford to pay good prices for well-bred seeds. We need as a national ideal, instead of an old age of idleness as a retired farmer in town, an old age of helpfulness in breeding for his fellow-man" richer grains, better fruits, and fairer flowers."

Plants lend themselves more readily to breeding experiments than do animals; for greater numbers of individuals may be produced at small cost, we may propagate or multiply the superior ones more readily, and in many instances, when we once get a superior kind, we can increase the number by stem propagation with little, if any, tendency to variation. One who is a master of the Mendelian laws may propagate from seeds with little danger of undesirable variation.

Plant Breeding Psychological

Then, too, the plant breeding is the psychologically correct place to begin agricultural study for it enables us to commence with something that the child can comprehend, and thus we may obey the law of apperception. Plant breeding enables the child to do something and to acquire something worth while, and thus it appeals to deep and lasting instincts of activity, acquisitiveness, imitation, and manipulation. It also enables us to bring into function the child's desire to get something more useful or more beautiful, and it appeals powerfully to curiosity and self-expression. Best of all, plant breeding enables us to give country people, both old and young, something to live for. Country boys and girls frequently lack ideals and purpose; plant breeding offers both. But we must not overlook the purely scientific side of plant breeding. The secrets of the universe are being drawn from nature to-day by the men who are in the plant-breeding work. Mendelism, the origin of sex, the possibilities of mutation, the cause of variation, and God's order for growing things are all to be found in the possibilities of plant breeding.

Name Improved Varieties

Local pride may beinculcated by having pupils name superior varieties should any occur. We need more farmers who, like Washington, put their names or the name of the farm or variety on the sack or box. " Made in Germany " is a label that is adding money to the pockets of the German people. Why may not " Grown in Pennsylvania, Iowa, Wisconsin or Minnesota " be made to add to the value of the products of the people of states using such labels ? This should be especially true of the products of farmers whose children have attended one of the redirected schools. Town people will readily give more for superior vegetables or fruit if they have a way of knowing that the product is really superior.

Set Standards High

God takes three months to make a pumpkin but Re takes one hundred years to make an oak." So it is in life—good things come slowly. It is a great thing for a teacher to set the standard as high as she is able to get her people to respond to. The Ohio farmers set their standard for their registered corn to the requirement that an ear be of known parentage and rank not less than fourth for yield, protein or whatever characteristic is claimed for it, and rank so in competition with not less than twenty-five other ears in row tests. The boys in the Evergreen Corn Club in Iowa took their scrub corn to a state show. This was before they learned what good corn is. Their corn looked so scrubby in comparison with the better corn at Ames, that for the first show the boys did not open their boxes. But one of them bought good corn for seed. They went home and they studied corn and they planted and cultivated their best. In three years they had to hire a railroad car to carry home the prizes which they won, and for the next few years they won practically every time. If a boy or man could be induced to take five years to produce a new corn, oats, potato or what not, and if he were wise enough to judge what is wanted, and then if he had the executive ability to get results, he could ask a high price and yet get buyers for all that he could produce. Some people make the bad blunder of advertising their product before they have multiplied it often enough. One man began too early to advertise an improved alfalfa with the result that he could not fill a tenth of his orders. Some man who understood what business really is was able in a year or so to fill the orders that should have gone to the plant breeder.

Begin Plant Breeding with the Best

It costs too much of human time and labour, and it costs too much of human life from the disheartening feelings at seed time, to allow land to be planted with inferior seed. The first step in improving our crops—whether vegetables, fruit, grains (as corn, rye, oats, wheat), grasses or potatoes and, for convenience, we may include cotton and tobacco—is to find what varieties are best for the district or section where the school is located. This does not mean that the teacher must necessarily confine her pupils to any one or two varieties.

The work may easily become too burden-some if there are too many varieties on the school plots, but if the pupils take up the work as home-project undertakings, then each may, after careful study as to which is best, select the one he and his parents decide is best. It will be time and energy wasted to start with inferior varieties; none but the best should be selected. Our Boys' and Girls' Club motto is, " To make the better best and the best better."

The State Experiment Stations carry on many experiments to determine the relative value of the different varieties of grains and vegetables. Hence, if the teacher wishes to know which are the best varieties for her locality, she may write to the experiment station for the information, or she may talk with some of the leading farmers of her district.

At this point she should remind the pupils of DeCondolle's law that plants should not be moved more than one degree north or south at any given time unless we wish to change their character. Webber has found that if climbing or twining beans or viny peas are transplanted from a southern to a northern climate, or from a lower to a higher altitude, they tend to produce dwarf types which show little tendency to twine.

Corn from Kansas, if planted in Iowa, seems to grow larger stalks and smaller ears. Farmers of Pennsylvania take advantage of these well-known phenomena and buy their seed for silo corn from the south, near Richmond, Va.

County Farms as Experiment Stations

In many places the county farm is used to determine the relative yielding tendencies of different crops. In Iowa, for example, Professor Holden had his helpers go over the county and get little paper sacks of seed corn from different farmers' planter boxes. This corn from each farmer was planted in a row on the county farm. The same was done with oats and wheat. When the crops were ready, the farmers came to the farm for a fall picnic, and there were able to see the relative yielding power of their seeds compared with that of others. One man found that his corn yielded thirty bushels to the acre, planted. side by side with that of a neighbor, whose crop yielded at the rate of seventy bushels to the acre. This work alone has enabled teachers to enrich their districts a number of times over the cost of running the schools. Page County, Iowa, where Miss Jessie Field introduced this work, sells over a million dollars of seed corn each year. Wright County, Iowa, where Mr. O. H. Benson while county superintendent introduced similar work and where the Evergreen Corn Club boys live, also receives many thousands of dollars for pure bred seed corn and other grains. There will be few with forethought enough to appreciate the value of the work at first, but, as Shakespeare says, " There be one or two whose good opinion outweighs the hundred." Then, too, teachers of intellect and foresight must get their approval for what they undertake largely from within their own consciousness.

Chapter 5

Microbiology in Biotechnology

Microbiology in Biotechnology

Activities of microorganisms are very important to almost every sector of concern to mankind. From a perusal of the foregoing topics, one can find applications (uses) of microorganisms to agriculture, forestry, food, industry, medicine, and environment. The scope and significance of microbiology has enlarged manifold, particularly when importance of environment was realised globally and the word environment was used in a much wider sense in terms of totality to include almost everything, every bit of nature. Demographic pressure on land with subsequent rise in demand for food and shelter had, out of several, two apparent impacts on nature and its resources -

(i) contraction (leading to deplection and in some cases finally to extinction) of biodiversity (fauna, flora and microbes), and

(ii) degradation of environment. Land became denuded, cultivable land area squeezed, air and water turned to be unfit for man, and wastes from urban, municipal and industrial areas piled up all around. Natural resources started depleting and there developed an energy crisis. It is in this context that the role of microbes was realised in order to exploit their vast potential to possible solution of some such problems.

New strains of microbes have been developed having the desired activities (entirely a new activity or enhancement activity in terms of rate productions). Environmental microbiology, a new field, could thus become of much relevance to us.

Most of these activities of microbes to agriculture, forestry, industry, medicine, food and environment have already been referred.

In this chapter an overview of this subject has been presented with the simple purpose of just a quick revision to remind the importance of microorganisms to human race

Microbes and Agriculture

Microbes and Agriculture - Besides being important in biogeochemical cycling of nutrients, microbes play vital role in maintenance of soil fertility and in crop protection.

Soil Fertility

Soil Fertility - Microbes are being exploited in two important ways - biofertilisers, and creating new nitrogen-fixing organisms.

Biofertilisers

Potential of Rhizobium, Azotobacter, Beijerinckia, Azospirillum, Cyanobacteria, such as species of Aulosira, Anabaena, Nostoc, Plectonema, Scytonema, Tolypothrix, and Azalia as biofertilisers has been exploited so as these could serve an alternative to chemical fertilisers.

Many brands of rhizobial inoculants are already in market today in the country. Several organisation and manufacturers are producing huge quantities of Rhizobium culture in the country.

These include Micro Bac., India, Shyam nagar, Parganas; Bacifil Inoculants, Lucknow; Govt. of Tamil Nadu; Nitro Fix Industries, Calcutta (W. Bengal) and Indian Organic Chemical Ltd. Bombay. In some other States, units are being prepared for increase in its production. Much progress has also been made with cyanobacteria in this direction.

Mycorrhizae, both ecto and endomycorrhiza help in uptake of N, P, K and Ca. They, particularly help in phosphorous nutrition.

Nitrogen Fixers

Nitrogen Fixers - Through recombinant DNA technology efforts have been made to introduce nitrogen-fixing genes (nif genes) into wheat, corn, rice, etc. Plasmids of the bacterium, E. coli and yeast are being worked out for such a possibility. Hybrid E. coli plasmid cloned with nif genes of a nitrogen-fixing bacterium, Klebsiella pneumoniae and hybrid yeast plasmids are then integrated.

Biopesticides

Biopesticides - Several microbes (viruses bacteria, and fungi) are being developed as suitable biopesticides for management of insect and nematodal pests. Some fungi have good potential of their use as bionematicides to control nema-todal pests of vegetables, fruit and cereal crops. Some bacterial and fungal products are also in use to control diseases of roots and shoots of plants.

Bioweedicides

Bioweedicides - Several fungi have been found very useful in the control of troublesome weeds of crop fields. Registered products are available in market for use in several countries.

Microbes And Public Health

Microbes And Public Health - The most successful bioinsecticide has so far been the bacterium, Bacillus thuringiensis. Several registered products of different strains of this microbe are on sale, thuricide being one, for the control of insects including mosquito-the carrier of malaria.

Microbes And Food

Microbes And Food- Applications of microbes to food, energy, industry and environment are interrelated in most situations. To cite an example, degradation of urban, municipal and industrial wastes by using a suitable (may be a tailored, genetically-engineered) strain of a microorganism should result into (i) dis-posal of pollutant, (ii) biotransformation of a waste- into a byproduct, suitable for consumption as food (conversion of agricultural waste into single cell protein is an example) and (iii) production of energy during such conversions. Therefore it is a three-way beneficial process carried out by a microbe.

Several yeasts, Saccharomyces, Candida, Torulopsis can be grown on waste materials, recycling them into food. Crude petroleum products are converted to SCP by Methylophilus methylotrophus. Cattle dung is fermented by methanogenic bacteria to biogas, an ideal biofuel in rural areas. Mushroom cultivation is good example of SCP where agricultural waste is recycled into food (SCP) and the left over residue used as organic manure.

Microbes And Industry

Microbes And Industry - Applications of microbes in industry are well known. Various microorganisms are used for commercial production of alcohols, acids, fermented foods, vitamins, medicines, enzymes etc. One recent development in industrial microbiology has been the production of immobilised enzymes and cells for production of these chemicals at enhanced rates with simultaneous recovery of the enzyme(s) involved in such processes.

New strains of microbes have also been developed through recombinant DNA technology for overproduction of metabolites. Immobilised enzymes and cells could have their maximum application in industrial microbiology. Immobilised enzymes have also been utilised in medicine.

Acetone Butanol Fermentation

Acetone Butanol Fermentations -The acetone butanol fermentation is one of the oldest fermentation known. The fermentation is based on culturing various strains of Clostridia in carbohydrate rich media under anaerobic conditions to yield butanol and acetone.

Clostridium acetobutylicum is the organism of choice in the production of these organic solvents. These fermentations were out of favour till very recently because of the availability of acetone and butanol from the petroleum industry.

Today there is considerable amount of interest in these fermentations. However, the concentration of end products in these fermentations is quite small and the fermentations are a type of mixed fermentation yielding a mixture of compounds such as butyric acid, butanol, acetone etc. Attempts to increase yeilds by use of genetidally altered strains or change in fermentation conditions have been partially successful ocassional careers, carried out in a professional way.

Microbes In Recovery of Metals

Microbes In Recovery of Metals - Recently microbes have been found very useful in enhanced recovery of metals including uranium from low-grade ores. Through bioteaching these microbes are able to solubilise the metals from their ores. Microbes thus play important role in mining and recovery of metals. For instance, Thio-bacillus thiooxidans and T. ferrooxidans can be used in recovery of copper.

Biotechnology in Agriculture : The New Green Revolution

The New Green Revolution

The green revolution which gave us plenty of grains to feed millions of people and the revolution in medicine which increased the life span of man are common knowledge even to lay people. All this was possible due to major discoveries and technological innovations in agriculture and medicine. Today, we are witnessing another revolution in biosciences be-cause of some major advances in cell biology and genetics.

Many people are inclined to believe that while the battle for green revolution of the type we saw three decades ago was fought in the field, the battle for the new revolution in biosciences known as "Biotechnology" is being fought in modem laboratories.

Some argue that the new biotechnology in agriculture is the second green revolution (part II) which will speed up crop improve-ment by gene manipulation in a Petri dish rather than in an open field.

The meaning or the definition of the word biotechnology has been the subject of hot debate by scientists and technocrats. The definition depends on the extent of expertise a group or an individual has with regard to cell biology. It also depends on the needs of a society or a country one lives in. The use of microorganisms or their products for food, feed, biofertilizers, biopesticides and medicine was known during the last 60 years.

The major developments in medicine and industry during those years came with the use of microorganisms for the benefit of mankind through new inventions in microbial technology and fermentation processes but the biotechnology of which we are currently talking about is envisaged by splicing genes and altering genomes by insertion of foreign DNA by genetic recombinant DNA techniques (the so-called genetic engineering).

The cell which is being manipulated by genetic recombination may be a microbial cell or a cell from a tissue of a plant, animal or man with the ultimate objective of inserting or cloning useful genes to obviate the use of long and tedious process of conventional breeding often replacing it by tissue culture techniques.

David Baltimore, Nobel Laureate, formerly Director, Whitehead Institute and Professor, Department of Biology, Mas-sachusetts Institute of Technology defined Biotechnology as the applica-tion of scientific and engineering principles to the processing of materials by biological agents to provide goods and services. The author defines biotechnology as the application of science and technology to accelerate or improve nature's processes in producing man's ever increasing needs for a good living. The path from green revolution to gene revolution or from conventional plant breeding to genetic engineering has been filled with many significant findings. For almost a century plant breeders identified and selected desirable characters and combined them into one individual plant. Since all characters are controlled by genes in chromosomes, plant breeding may be regarded as manipulation of chromosomes.

This was done by the sorting and retention of similar chromosomes in the same plant to reach a homozygous state, a method termed pure line selection. Alternatively, different chromosomes can be combined to form a heterozygous state, a method known as hybridization conferring hybrid vigour or heterosis.

The next step was the development of genetic variability through spontaneous or artificially induced mutations. Normal plants are diploids but when plants are developed with three or more sets of chromosomes, they become polyploids that' tend to be bigger than diploids. Autopolyploid plants have genes similar to their diploid ancestors whereas allopolyploids are combinations of genomes of two different species that differ in characteristics.

The first achievement of hybridization techniques was the development of hybrid maize in 1919 which revolutionizep American agriculture. The development of hybrid wheat and rice plants in 1960s filled the bread basket of developing countries, generally known as green revolution, for which Dr. Norman Borlaug was awarded the Noble Peace Prize in 1970.

The discovery that plant cells can develop into entire plants was another land mark in the development of new varieties of crop plants. The term tissue culture was coined to denote the in vitro development of plants in test tubes from calluses generated from plant parts. This led to mass production of uniform plants and revolutionized floriculture in the globe. Tissue culturing often leads to progenies which are variable. These progenies are known as somoclonal variations and have been exploited to generate mutations and it has been estimated that in vitro tissue cultures can produce ten times more somoclonal variations than that can be induced by chemical mutagens.

Fusion of naked genetically compatible protoplasts (inter specific) resulted in successful regeneration of new varieties. Fusions between incompatible protoplasts (inter generic) resulted in abortive cell division and successes in regeneration was never achieved, excepting the instance of crossing between tomato and potatoes forming 'pomatoes' which can only be regarded as a laboratory success not amenable to commercial exploitation. With the advent of biotechnology, agriculture has reached a science based industrial state. By using recombinant DNA technology, many transgenic life forms have been engineered since 1985. Transgenic plants belonging to both monocotyledonous plants such as maize, millet, wheat, rice and ragi and dicotyledonous plants such as alfalfa, clover, peas, soybean, mothbean, potato, tobacco, cotton, flax, sugarbeet and sunflower have been constructed. New varieties of vegetables and fruits such as cabbage, carrot, cauliflower, celery, cucumber, horseradish, lettuce, rape, grape, muskmelon and strawberry have been developed. The new varieties have incorporated genes capable of resisting one or more of the following: herbicides, insects, stress, frost or virus infections.

Plant biotechnology has opened up the possibility of producing artificial seeds, artificial sweetners (sugar substitutes) and bioplastics. Normal seeds have an embryo surrounded by cotyledons for initial sustenance during germination.

By somatic hybridization, plant embryos can be mass multiplied in fermentation tanks and each embryo is then encapsulated in a jelly-like coat that can be called an artificial seed. Some estimates have revealed the possibility of production of 80,000 embryos per day but the cost could well be prohibitive for commercial exploitation. Presently, several companies are engaged in reducing the cost for atleast some crops such as carrots and celery.

The most important sugar substitute is the maize based high fructose com syrup (HFCS) known as isoglucose in Europe. Some estimates put HFCS production worldwide to 6 million tonnes available in liquid as well as crystal form.

Aspartame is a synthetic chemical thousand times sweeter than sugar. With the advent of this product nearly 38 research institutes and companies around the world are engaged in producing novel chemical sweetners.

Thaumatococcus danielli or commonly known as Katemfe is a plant that grows in humid forests in Western and Central Africa. The berries of this plant contain the protein thaumatin that is 2500 times sweeter than sugar. Tate and lyle, a UK based sugar company had set up plantations of Katemfe in Ghana, Liberia and Malaysia. The frozen berries were processed to obtain purified thaumatin that was sold under the brand name 'Talin'. One drawback of thaumatin is its lingering taste limiting its use in food products. In spite of this, research is underway to understand the gene coding for thaumatin and its transfer to E. coli and other plants.

Stevia rebaudiana grows in Paraguay and several countries in South East Asia. The plant is capable of producing proteins several hundred times sweeter than sugar. The product is being marketed in Japan which has also bitter taste. African forests abound in sweet berries such as 'Miraculous berry' (3000 times sweeter than sugar) and Mexico has Lippia dulcis, thousand times sweeter than sugar. The search for cheaper substitute to cane sugar is being pursued vigorously and in future years we may have alternate sugar sources.

An interesting example of how a plant can be made to produce novel chemicals such as bioplastics is the transgenic Arabidopsis capable of producing granules of polyhydroxybutyrate (PHB), a polyester which is normally obtained from the bacterium Acaligenes eutrophus. In fact, PHB is a storage product in many bacteria intended to be used as a source of energy by bacterial cells in times of nutritional stress.

This bioplastic material is a delicate product destroyed by pH above 8 and temperatures above 70°C. The product is mixed with polyhydroxyvalerate (PHV) to make it flexible and moulded into any shape, spun as fibre or rendered into a film. It is biodegradable to CO_2 and H_2O with no environmental hazard. The bioplastic is compatible with living tissue and hence can be adapted for medical purposes. It can also be used as a much in agriculture.

Transgenic animals and microorganisms are being used for fundamental research, for production of pharmaceuticals such as goats-Iactoferrin and for production of biological control agents. Pigs and rabbits are genetically engineered to function as organ donors for human beings and chickens have been exploited for producing foreign proteins in eggs.

Pharm biotechnology (agricultural production of pharmaceutical products) has to be distinguished from Farm biotechnology (productivity related agricultural applications) and very likely agriculturally produced pharmaceuticals will be marketed sooner than agricultural products, despite the fact that the latter could undoubtedly increase global food supply.

This has been due to success in pharmaceutically oriented animal experiments as exemplified by the transgenic modification of pigs and sheep for expressing valuable pharmaceutical products in milk where all animals in the offspring appeared healthy contrasting with the transgenic pigs generated to produce leaner meat or more rapid growth whose offspring had adverse effects.

There are about 20 different man made pharmaceuticals involving crops that have been genetically changed to produce a range of prophylactics from cholera vaccines, herpes vaccine and cancer treatments.

Potatoes seem to be ideal vehicles for the new generation of vaccines such as vaccine against E. coli disorders of the intestine. These are friendly and easier to tolerate than injections.

Phosphorus in seeds is a poor nutrient for monogastric animals such as chickens unless phytase is present to release phosphorus. Feeding chickens with seeds containing the phytase gene from Asperigillus niger brought about growth increases in chicken. This biotechnological innovation known as "gene farming" not only improved the quality of chicken feed but also minimised the excretion of phosphate in the environment. Another example is the case of sweet potato which is a staple food in China. The strategy here was to implant twin genes such as viral coat protein gene and Bacillus thuringiensis genes into sweet potato to ward off diseases caused by viruses as well as insect pathogens.

Nucleic Acids

It would be helpful to briefly describe some basics of molecular biology before attempting to understand its implications in biotechnology. The classical discovery of the structure of genetic material by Watson and Crick in 1953 revealed the unique suitability of nucleic acids for carrying genetic information and transmitting to subsequent generations.

All the information needed for growth and multiplication of most organisms is carried by nucleic acids, especially the double-stranded deoxyribonucleic acid (DNA) or single or double stranded ribonucleic acid (RNA).

RNA differs from DNA in that the single strands have a ribose instead of deoxyribose and uracil in place of thymine. The double stranded DNA occurs in a helical structure with a backbone of alternating phosphate and deoxyribose molecules having a purine or a pyrimidine base linked to the I-position of each sugar molecule. The two complementary strands of DNA are twined together by hydrogen bonds between the purine and pyrimidine base pairs: adenine-thymine (AT) and guanine and cytosine (CT). An adenine in one strand of DNA occurs directly across a thymine in the other strand. Similarly, a guanine (G) occurring in one strand is bonded to a cytosinc (q across the other strand).

The genetic information is coded by the linear arrangement of bases on the DNA strands. The sequence of nucleotidase dictates all the characteristics of an organism and serves as a genetic code. The sequence of nulceotides, read in groups of three (triplet) reflects the sequence of amino acids in the large number of proteins (enzymes) synthesised by a cell. Each triplet is known as a codon and there are 64 possible combinations beginning with four nucleotides.

Protein Synthesis

Mitosis (cell division) results in the formation of tissues. Before cell division takes place, the chromosomes in the parent cell have to be duplicated so as to be equally shared by the daughter cells. This replication is carried out within the nucleus of the cell by the action of an enzyme known as DNA polymerase.

The synthesis of proteins takes place within the cytoplasm of the cell away from the nucleus. The genetic information in the DNA is conveyed to ribonucleic acid in ribosomes residing in the cytoplasm by a process known as transcription. During transcription, only one of the two strands of DNA becomes translated into RNA by RNA polymerase. The synthesis of RNA always proceeds in a fixed direction beginning at the 5' end and concluding with 3' ended nucleotide.

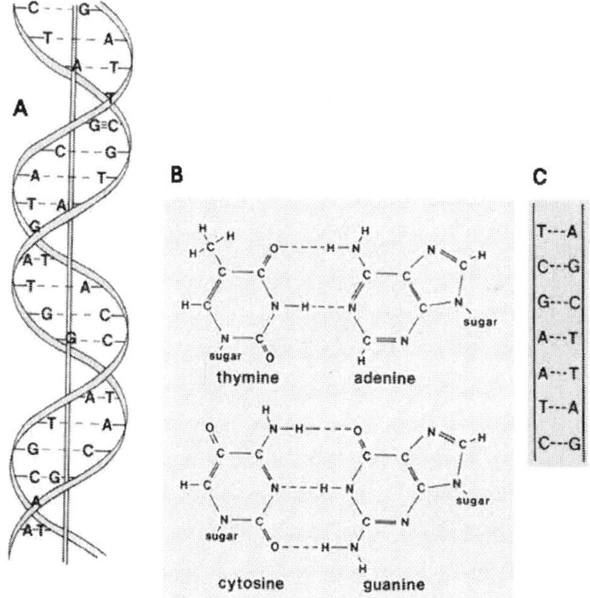

Figure: *The double helix. a. Deoxyribonucleic acid (DNA) is a double-stranded, helical molecule composed of nucleotide units. b. Diagram of the molecular structure of the base pairs of DNA. c. Diagram showing pairing of the nucleotide bases in a short DNA segment.*

The synthesized RNA from DNA of the nucleus in a cell moves into the ribosomes of the cytoplasm carrying with it information needed to synthesize protein in the ribosome. Hence, this RNA has come to be known as messenger RNA (mRNA). The 5' end of an mRNA molecule attaches to a ribosome.

		U		C		A		G	
	UUU	Phe	UCU		UAU	Tyr	UGU	Cys	U
U	UUC		UCC		UAC		UGC		C
	UUA	Leu	UCA	Ser	UAA	Stop	UGA	Stop	A
	UUG		UCG		UAG	Stop	UGG	Trp	G
	CUU		CCU		CAU	His	CGU		U
C	CUC		CCC		CAC		CGC	Arg	C
	CUA	Leu	CCA	Pro	CAA	Gin	CGA		A
	CUG		CCG		CAG		CGG		G
	AUU		ACU		AAU		AGU		U
A	AUC	Ile	ACC	Thr	AAC	Asn	AGC	Ser	C
	AUA		ACA		AAA		AGA		A
	AUG	Met	ACG		AAG	Lys	AGG	Arg	G
	GUU		GCU		GAU		GGU		U
G	GUC		GCC		GAC	Asp	GGC	Gly	C
	GUA	Val	GCA	Ala	GAA		GGA		A
	GUG		GCG		GAG	Ghu	GGG		G

About 4 per cent of total cellular RNA is mRNA. Since only a small segment of mRNA is attached at a given time to a ribosome which is moving across it, a single mRNA molecule can be read at the same time by several ribosomes, occurring as polyribosomes consisting of anywhere from 6 to 50 ribosome units.

Before the process of amino acid polymerization into proteins begins, the 20 different amino acids in a cell are first transformed into energy-rich precursors. These precursors get attached to a small transfer RNA (tRNA) molecule. A group of three nucleotidase (triplet) constitute a tRNA and serves as an codon that uses base pairing to find three nearby nucleotidase which in turn serves as a codon on a mRNA molecule.

Specific enzymes known as aminoacyl synthetases now begin to bind or attach aminoacids to specific tRNA molecules. Ribosomes are thus minifactories for protein synthesis by a series of codon-anticodon interactions which occur on their surfaces.

These interactions take place when mRNA molecules move across the active surface of ribosomes aligning successive codons into position to form successive aminoacids along polypeptide chains and this process is known as translation.

There are 64 potential codons and of these 61 are used to specify amino acids while 3 are made use of to provide signals to terminate the formation of polypeptide chains. Several amino acids are determined by more than one codon. All the codons put together constitute the genetic code and this code is common to all forms of life including microorganisms.

Southern and Northern Blot Techniques

Southern blot is the classic technique described by Southern in 1975 for understanding individual genes in a complex mixture of DNA.

The pro-cedure involves cutting up high molecular weight genomic DNA into fragments by enzymatic digestion with restriction endonucleases. These bacterial enzymes have been isolated and are available in many laboratories.

They have been prepared from several bacteria and are designed as molecular scissors which cut the strands of DNA at specific oligonuleotide recognition sequences. The fragments of DNA are then separated on the basis of size by agarose gel electrophoresis. The fragments with smaller number of nucleotides move faster on the gel in an electric current than the larger ones which have higher number of nucleotides.

At the end of the experiment, the gel is immersed in a fluorescent dye (ethidium bromide) which binds to DNA. When viewed under ultra-violet light, the fragments can be visualized as separate entities reflecting the nucleic acid pattern. The cellular RNA can also be run on agarose gel and the fragments can be separated in much the same way as DNA. This technique is known as Northern blot technique.

Recombinant DNA Technology and Gene Cloning

Man made nucleotide sequences tailored to carry desired traits, known as molecular probes have been extensively used in molecular biology studies. These nucleotide sequences (probes) ought to be correct, pure and available in large amounts. The probe must be labelled in some manner to help detection. These probes are prepared by recombinant DNA techniques using the common enteric bacterium *Escherichia coli.* This organism has been studied so often by molecular biologists that they are now aware of the implications of its genome containing about 4.2 million base pairs (bp).

Furthermore, the bacterium has extrachromosomal DNA molecules known as plasmids that can replicate autonomously independent of the nuclear DNA. These plasmids are closed single pieces of supercoiled DNA which can be stably inherited by daughter cells. They have the ability to replicate to high numbers (copies) within each bacterial cell.

The restriction endonucleases cleave DNA only to specific oligonucleotide sequences rather asymetrically leaving 'sticky' ends. The sticky ends remain complementary between any two DNA fragments sliced by the same restriction enzyme. This property makes it easy to insert or 'clone' an outsider or foreign gene to the E. *coli* plasmid provided it has been sliced by the same restriction enzyme.

The transformed plasmids known as 'vectors' can be amplified to a high copy number in a standard bacterial culture. To retrieve the inserted foreign DNA, the bacterial biomass from the culture medium is separated and lysed. The DNA content is purified and sliced by the same restriction enzyme that was earlier used for original cloning.

Plasmid vectors have been used to transfer DNA from one prokaryotic cell to another, from a prokaryote to an eukaryote and from an eukaryote to a prokarotic cell. The plasmid vectors have the limitations of cloning upto 5000 base pairs (bp) or 5 kilobases (kb). By developing and using bacteriophage lambda chromosomes, foreign DNA have been cloned upto 15 kb.

The ability to use vectors has been further enlarged by using features of both plasmids and bacteriophage lambda to the extent of 50 kb. Presently, specially constructed DNA fragments from yeast cells known as yeast artificial chromosomes (YAKs) are available that can be used to clone pieces of DNA upto 1 million bp.

One of the important steps in cloning a foreign gene is to obtain the desired gene in the absolute pure condition. This can be done by a traditional method beginning with the purified protein which the gene produces. Required antibodies are raised which will recognize and precipitate the protein when added to a cell extract from a tissue where the protein is actually synthesized.

This results in the precipitation of newly made polypeptides which are being elongated on the polyribosomes. However, the precipitate contains unwanted ribosomes mixed with the mRNA templates required for producing the protein in question. When these mRNA templates are purified from the mixture, a sequence of nucleotides complementary to the gene of interest can be obtained. By using a retroviral enzyme reverse transcriptase, which synthesizes DNA from a RNA template, a cDNA sequence from the mRNA sequence can be obtained.

By the addition of DNA polymerase the second strand of DNA can be replicated which results in the formation of a double stranded DNA copy of the gene of interest. This elaborate procedure is known as the polysome precipitation.

In recent years, polysome precipitation method has been replaced by 'DNA libraries'. A cDNA library is made up of all the actively transcribed genes of a tissue inserted into a population of bacterial cells. The bulk of mRNA preparation is reverse transcribed and inserted into plasmids in one lot, with the objective that every possible cDNA sequence will be carried by atleast one bacterial cell in the culture. This cDNA library has to be sorted out for a particular cDNA sequence of interest.

The bacterial culture containing the cDNA is spread out on agar plates and filter blot technique is used with a labelled oligonucleotide probe (1D-40 bp) to select a colony containing the foreign gene of interest. Presently, many biotechnological companies have developed automated instruments that can rapidly synthesize oligonucleotides of any sequence.

Gene Exchange in Bacteria

Gene e change in bacteria takes place by processes known as transformation, transduction and conjugation. The addition of foreign DNA to actively growing bacterial culture results in chance entry of foreign DNA into cells by modification of bacterial cell envelope followed by the intake of DNA into the bacterial genome. Transduction involves bacteriophages or bacterial viruses whose DNA enters the bacterial cell followed by the disintegration of the bacterial chromosome. The phage DNA multiplies in the cells, the cell walls undergo lysis releasing the phages which have multiplied in the mean time.

The phages which mediate this type of transduction are known as lytic or virulent phages. On the contrary when a temperate phage (non-virulent type) enters the bacterium, the phage DNA becomes attached to the bacterial chromosome and remains integrated with the bacterial genome for many generations. This process is also known as lysogeny. At times, temperate phages may turn virulent leading to lysis and production of more bacteriophages. The temperate phage, when freed from the cell may carry with it small pieces of DNA which upon delivery to a next host cell can add an additional character to the new cell's capabilities.

Contact between two cells is required for conjugation, one acting as a donor and the other a recipient. The donor possesses a fertility factor (F^+) and the recipient has no such factor (F^-) but must be viable for successful conjugation.

The Agrobacterium Mediated Transfer of Genes

A naturally occurring conjugation phenomenon in *Agrobacterium tumefaciens* induced crown-gall disease of plants has been ably exploited by biotechnologists to genetically engineer foreign genes into dicotyledonous plants rendering them transgenic in characters such as resistance to viral diseases, herbicides or bioinsecticides.

A. Bacterial transduction with bacteriophages. A virulent phage

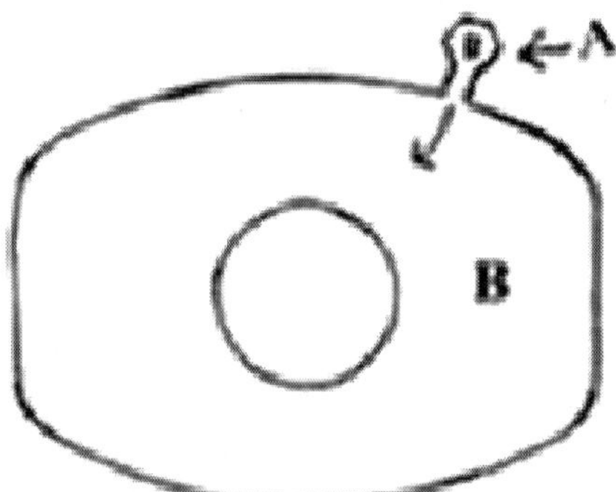

C. The phage DNA is injected into the bacterial cell followed by disruption of bacterial DNA and the phage takes Control of bacterial cell functions

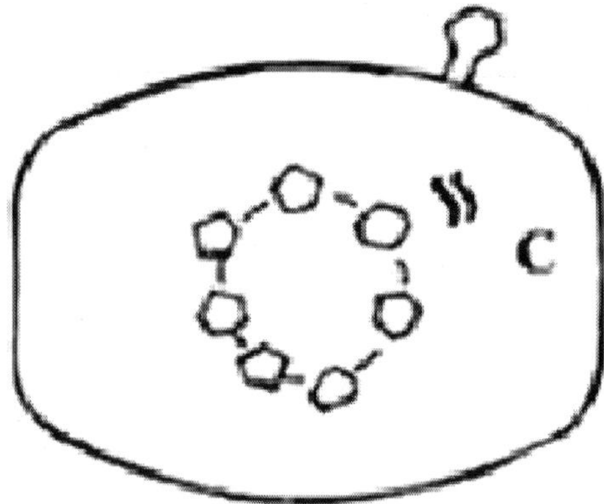

D. New bacteriophages are formed and released by cell lysis

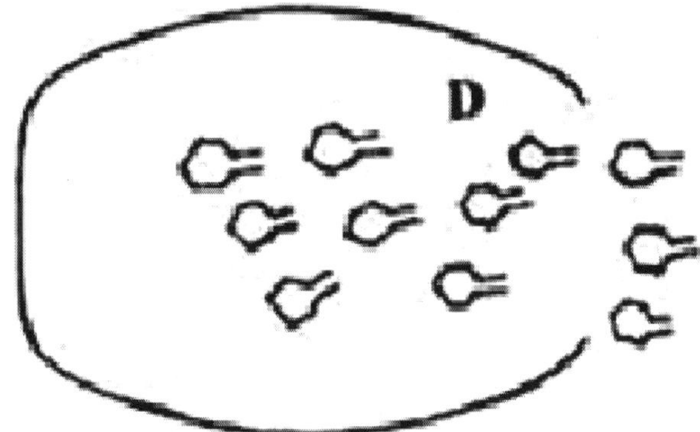

E. In a phenomenon called lysogeny infection with a temperate phage results in the integration of phage DNA with the bacterial Chromosome

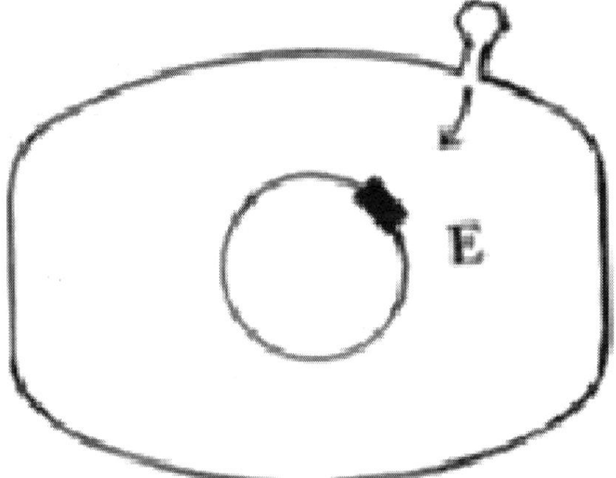

F. Maintained as such for several generations. occasionally phage DNA gets detached from the bacterial chromosome

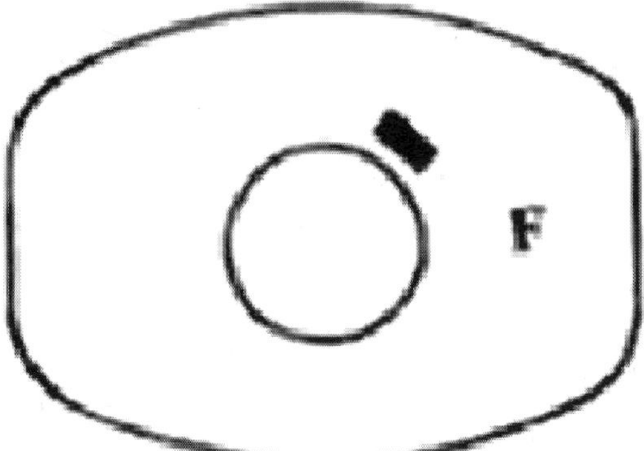

G. The bacterial DNA and takes control of the bacterial cell behaving like a virulent phage causing lysis of the cell with the release of mature bacteriophages

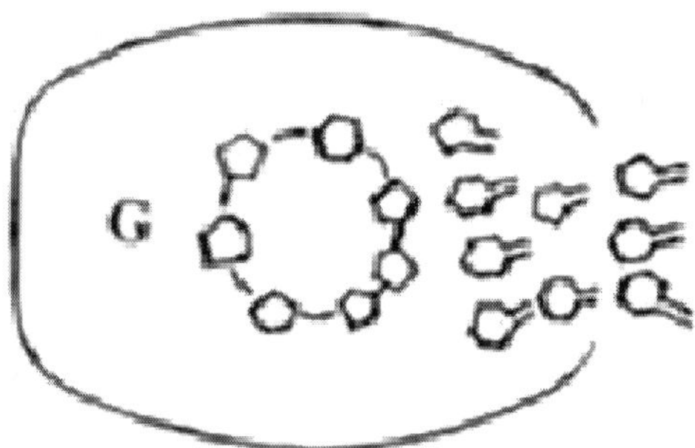

At is a soil bacterium which infects the crown region of dicotyledonous plants (monocots are resistant) through mechanical wounds to produce crown galls. A.t harbours large extrachromosomal megaplasmids. Most of the genes required for tumour formation are located on one such 180 kb megaplasmid designated as Ti plasmid, the letters Ti denoting the tumour inducing ability of the plasmid. The Ti plasmid contains a tumour inducing region (T-DNA) which also carries genes for the synthesis of two growth hormones, the IAA and cytokinins and the genes controlling the synthesis of a group of amino acid derivatives known as opines (nopaline and octopine).

The expression of bacterial genes con-trolling growth hormone production is not controlled by the plant but is necessary for the-development of crown gall symptoms. The opines synthesized serve as nitrogen and carbon sources for the bacterium. However, the genes controlling the growth hormone production and opine synthesis are not essential for transferring or integration of T-DNA into the host plant cell genome.

The Ti plasmid has also a cluster of about 8 genes known as virulence genes *(vir* genes). This gene cluster of 35 Kb DNA is necessary for the recognition of susceptible cells on the plant surface, to excise T-DNA from the plasmid and transfer the T-DNA region to the host cell. The *Vir* genes are activated only by contact with cell metabolites released by the wounded plant and they are not functional or expressed in pure A.t cultures grown on synthetic media.

There are two border regions to T-DNA (LB and RB) which are known to contain genes involved in the secretion of the enzyme endonuclease that scissors off the T-strand from the Ti plasmid. Thus the genes encoded within the T-strand have all the appropriate signals for efficient transcription and translation in their eukaryotic host.

The process of infection begins with the bacterial surface components of A. t. recognizing the plant surface which is susceptible to the attack by the pathogen, followed by a process analogous to bacterial conjugation whereby a single strand of T-DNA (the T-strand) is transferred to the plant cell probably through pores in the cell wall.

Within the host cell, several copies of T-DNA are inserted at single or multiple sites in the host chromosomes and function as typical eukaryotic chromatin.

The earlier procedure to identify transformed cells was to develop plant galls or tumours from such transformed cells. Alternatively, the transformed cells were grown in hormone independent cultures. However, in later experiments, modified T-DNA and Ti plasmids have been used to facilitate rapid experimentation and development of transformed plants. In the modified T-DNA, the genes coding for phytohormones and opines have been excised because, as stated earlier, these genes are not concerned with the transfer and integration of T-strand into the host genome but are essential for the manifestation of crown-gall symptoms.

In essence, deleting these genes from T-DNA results in the elimination (disarming) of the oncogenic (tumourous) phenotype. Foreign DNA can be inserted within the right and left border regions of an appropriate plasmid, thereby enabling the delivery of large multigenic segments of DNA into plants. In this way T-DNA region has been engineered (disarmed) to eliminate the tumour inducing genes. Selection of transformed cells has been facilitated by engineering a variety of antibiotic markers selectable in plants such as resistance to kanamycin and gentamycin.

Ti plasmids have also been engineered in other Agrobacterium spp. and E. coli to facilitate efficient plasmid replication and selection. Such manipulated plasmid vectors could be used directly in transformation steps obviating the need to repeatedly infect plants with A. tumefaciens cells.

Procedures for Agrobacterium Mediated Transformations

Several procedures have been followed till to date: (a) through conventional gall formation on wounded plant stems or.leaves that are not amenable to monitoring with disarmed vectors, (b) Co-cultivation of *Agrobacterium* with plant protoplasts, a method that has been successful in plants where regeneration of plants from protoplasts has proved to be successful and (c) tissue transformation with ex-plants (leaf, cotyledon section or somatic embryos) inoculated with *Agrobacterium,* a procedure that has proved faster than protoplast regeneration.

When shoots are formed in regenerated plants, the plantlets are transferred to potted soil for acclimatization to natural surroundings.

The introduction of *nif* genes controlling nitrogen fixation in some microorganisms, into higher plants other than legumes is an experimental strategy envisaged by several research groups.

Direct DNA-Transfer Technologies

Most monocotyledons like cereals and sugarcane are not amenable to *Agrobacterium* technology because they are resistant to infection by the bacterium. Alternative technologies are being developed which include the introduction of foreign DNA into plant protoplasts mediated by polyethylene glycol and poly-L-ornithine, electroporation (use of electric current), calcium phosphate coprecipitation, liposome fusion, microinjection and particle bombardment. Regeneration of plants from transformed protoplasts is labour intensive and subject to the possibility of develop-ment of somoclonal variations.

The use of 'microprojectiles' or 'particle bombardment' is a procedure for delivering foreign DNA into plant cells. The technique involves coating small gold or tungsten beads with plasmid DNA and propelling the beads to intact cells, embryos or differentiated tissues using high velocity particle , guns' or electrical discharges. This technique has been useful in delivering DNA into the nucleus and mitochondria of yeast, nucleus and chloroplasts of *chlamydomonas,* epidermal tissues of *Allium cepa* and suspension cultures of maize. A major limitation of this technology arises from the paucity in the number of particle guns available and the high cost in building them.

Antisense RNA Strategy

The basic idea in antisense strategy is to block the expression of a particular gene product with the help of a transgenic construct containing the gene or part of the gene with the transcript in reverse orientation with respect to the promoter. This procedure results in the formation of a complementary RNA rather than the normal sense RNA. This antisense RNA binds with its homologous sense RNA, preventing translation and/ or facilitating degradation or accumulation of a gene product by 90 to 99 per cent, thus creating a phenotypic mutant. There have been many reports of insertion of antisense construct of polygalacturonase enzyme into tomato plants. The resulting transgenic tomato plants had low levels of this enzyme and hence their fruits ripened slowly thereby increasing shelf life. Antisense RNA technology has also been used with success towards minimizing viral diseases.

Frost Control Biotechnology

Frost injury to plants is caused at temperatures less than 0°c. Two kinds of frost injuries have been recognized: those occurring above minus 5°C and those occurring when temperature drops below-5°C. Plants resist frost by restricting freezing to intercellular spaces and adjusting the water potential intracellularly to reach equilibrium with the ice formed in inter-cellular spaces. Frost injury takes place when this equilibrium is upset and the rate of intracellular ice format on exceeds that of intercellular spaces followed by death due to disruption of cell membrane properties.

Frost tolerant plants, however, have endogenous ice-tolerance mechanisms.

How does frost sensitive plants overcome frost injury. There appears to be a super cooling mechanism in such plants to avoid ice formation. Ice nucleation or initiation of ice embryos is caused by the orientation of water molecules by organic and inorganic substances. Several plant-associated bacteria are highly active in ice nucleation and may be responsible for frost injury and the biotechnological implications of this activity have been studied with reference to frost injury at temperatures above minus 5°c.

Several strains of bacteria such as *Xanthomonas campestris, Pseudomonas viridiflava,* P. *fluorescens* and *Erwinia herbicola* inhabit the epidermal crevices and hairs of leaf surfaces and are active in ice nucleation at temperatures above minus 5°C. It should be noted that the efficiency of ice nucleation activity of these bacteria differs not only with the strains of bacteria but also with the plants harbouring them. The population density of ice nucleating bacteria is variable on plant surfaces depending upon the species of plants and the environmental conditions under which they grow. These epiphytic bacteria are known to occur on frost resistant as well as frost susceptible plants and are known to be killed by disinfectants and V.V. light.

The genes conferring ice nucleation characteristic have been partially characterized from P. *syringae,* P. *fluorescens* and E. *herbicola* and are known to be a single contiguous region of approximately 4000 bp. They have been cloned in *Escherichia coli.*

Streptomycin and oxytetracycline as well as *copper* hydroxide applications to leaf surfaces reduced the incidence of frost injury, despite the fact that dead cells were also known to be active as ice nucleating agents. Likewise, non-ice nucleation-active bacteria can also reduce the population of ice-nucleating ones both in the green house and the field by limiting nutrients to the ice nucleating bacteria. For example, non-ice nucleation-active (Ice mutants of P. *syringae* reduced the population size of ice nucleation active parental strains of P. *Sryringae* that were co-inoculated on pretreated plants.

Field trials have been conducted to understand the competition behaviour of ice" strains of P. *syringae* by inoculating these strains to potato plants. Ice' strains dominated the leaf surfaces for the first 4-6 weeks after inoculation. The population of Ice+ strains on plants colonized by Ice- P.*syringae* strains was significantly decreased in comparison with uninoculated plants.

The incidence of frost injury to potato plants inoculated with Ice strains was significantly lower than uninoculated control plants in natural field frosts in a field experiment in California.

The use of microorganisms for competitive control of frost injury have been found to be effective only when applied to young vegetative plants in the field because such young vegetation may not have been extensively colonized by other epiphytic microflora.

Minimal occurrence of extraneous nicroflora on leaf surfaces which can be achieved by the application of bactericides such as cupric hydroxide can result in micro-habitats most conducive for the functioning of Ice- P. *syringae* in large numbers on the leaf surface so that the Ice-strains can effectively compete for nutrition with frost inducing naturally occurring wild strains of P. *syringae.* In-tegrated chemical and biological control measures to contain frost injury to plants appear to be a desirable approach. In addition, co-application of copper resistant Ice strains of P. *syringae* and cupric hydroxide has also been considered as an attractive proposition to control frost injury.

Virus Resistance in Transgenic Plants

There have been many reports about the development of transgenic plants which have become resistant to virus infections. These plants showed resistance to virus infections when they were transformed with sequences related to several gene functions.

Some of the areas where successes have been seen related to sequences concerning viral capsid protein, viral move-ment protein, antisense RNA, antibody-mediated resistance, interferon-related genes and host genes involved in plant protection. Viruses are biochemical complexes consisting of a RNA or a DNA genome packaged into a protein capsid which mayor may not be sur-rounded by a membrane envelope. The protein coat 'covered genome is referred to as the nucleocapsid. The proteins on the surface of the capsid and envelope determine the interaction of the virus with the host and elicit the protective immune response against the virus. Some virus particles also contain enzymes required to facilitate the replication of the virus.

The tobacco mosaic virus (TMV) is an example of a virus with helical symmetry whose capsomeres (many protein subunits of the capsid) appear as projections that are assembled on the RNA genome into rods extending to the length of the genome. In other viruses the capsomere arrangement is cubical or icosahedral enclosing its nucleic acid component. Engineering resistance in plants involves either countering the capsid properties or disrupting the virus replicating mechanisms in the hot.

Transformation of Sequences Related to Viral Capsid Protein

Coat mediated resistance is the expression of a gene that causes the transformed cell and regenerated transgenic plants to produce the coat protein (CP) of a virus thereby conferring resistance to infection in the transgenic plant. Plants that accumulate large amounts of coat protein escape/mini- Elementary features of typical virus particles: A-Enveloped icosahedral virus; B-Naked icosahedral virus; C-Enveloped helical virus and D-Naked helical virus maize virus infection.

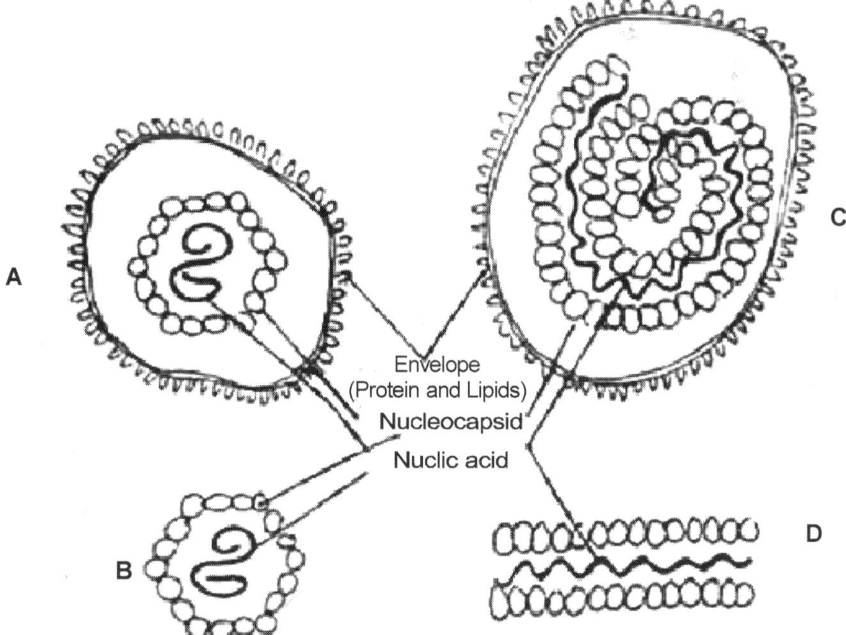

Figure: *Elementary features of typical virus particles.*

The CP mediated resistance was first reported in tobacco mosaic virus (TMV) in 1986. Subsequently several instances of resistances for several virus diseases of plants have been reported which include tomato mosaic virus, alfalfa mosaic virus, cucumber mosaic virus, tobacco streak virus, potato virus x and y, tobacco etch virus, tobacco rattle virus, potato leafroll virus, potato virus S.

A point which needs to be stressed is that CP mediated resistance does not provide immunity to transgenic plants; for instance TMV resistant plants that expressed CP gene were susceptible to inoculum levels of 10mg/ml while plants that were not transgenic got infected even at inoculum levels of 0.001-0.01 mg/ml.

Despite the fact that the mechanism involved in CP mediated resistance has not been clearly understood, some kind of inferences can be made from studies on TMV diseases in tomato and tobacco. Firstly, the transgenic plants are protected due to reduction in the number of infection sites that means fewer number of plants may get infected than controls. Secondly, the transgenic plants are less likely to develop systemic infection and thirdly, such plants may produce less virus particles than the ones not having the resistant gene.

Transformation of Sequences Related to Viral Movement Protein

A gene product of TMV was the first to be directly identified and investigated in viral movement. Plasmodesmata are the portals between adjacent host plant cells responsible for viral movement.

A gene product (viral protein) has been found to accumulate at or near plasmodesmata of TMV infected transgenic plants especially in the cell wall fraction, which directly or indirectly changes the plasmodesmatal permeability to allow the passage of TMV particles measuring 18 x 300 nm (30-KDa) while in the normal tobacco plants only low molecular weight compounds less than 1.5-2.0 nm could pass through.

However, microinjection of recombinant viral protein into plant cells resulted in the enhancement of plasmodes-matal expansion upto 9nm limit, that was insufficient for viral movement to take place between host cells and hence an alternate explanation had to be found.

It is known that the viral protein also binds to RNA and hence it has been conceived that one part of the viral protein binds to TMV virus RNA and shapes into a complex measuring about 2nm and the other part of the protein binds to plasmodesmata to increase their permeability character and thus enable the virus to pass through plasmodes-mata.

Resistance Conferred by Antisense RNA

Antisense regulation of gene expression is a natural phenomenon concerning the specific expression of a nucleotide strand which is negative to a certain gene transcript capable of intervening the expression of that gene at different levels. Antisense RNA may bind to the loop initiated by RNA polymerase in the nucleus and interfere in the initiation of transcription.

It may function in the cytoplasm by hybridizing with mRNA leading to translation arrest by preventing the binding of ribosomes to mRNA. Antisense regulation of gene expression has been successfully exploited in plants. Some examples are inhibition of flower pigmentation by antisesnse for chalcone synthase, intervention in the expression of ribulose biphosphate carboxylase in tobacco and polygalacturonase in tomato.

Most plant viruses replicate in the cytoplasm and do not go through any nuclear phase and therefore antisense RNA may either intervene in translation or promote mRNA degradation. Antisense mediated resistance has been achieved in transgenic plants against several plant viruses including TMV. This mode of resistance appears to be milder than coat—-mediated protection.

Antibody-Mediated Resistance

Antibody-mediated protection (akin to hybridoma-derived monoclonal antibodies) from only two plant viruses (TMV and artichoke mottle crinckle virus) are known. The mode of action of antibodies appears to be through neutralization of viral surface proteins in a way that intervenes the initial establishment of infection.

Chapter 6

Relation of Bacteria to Agriculture

The Silo

In the management of a silo the farmer has undoubtedly another great bacteriological problem. In the attempt to preserve his summer-grown food for the winter use of his animals, he is hindered by the activity of common bacteria. If the food is kept moist, it is sure to undergo decomposition and be ruined in a short time as animal food. The farmer finds it necessary, therefore, to dry some kinds of foods, like hay. While he can thus preserve some foods, others can not be so treated.

Much of the rank growth of the farm, like cornstalks, is good food while it is fresh, but is of little value when dried. The farmer has from experience and observation discovered a method of managing bacterial growth which enables him to avoid their ordinary evil effects. This is by the use of the silo. The silo is a large, heavily built box, which is open only at the top. In the silo the green food is packed tightly, and when full all access of air is excluded, except at its surface. Under these conditions the food remains moist, but nevertheless does not undergo its ordinary fermentations and ;putrefactions, and may be preserved for months without being ruined. The food in such a silo may be taken out months after it is packed, and will still be found to be in good condition for food. It is true that it has changed its character somewhat, but it is not decayed, and is eagerly eaten by cattle.

We are yet very ignorant of the nature of the changes which occur in the food while in the silo. The food is not preserved from fermentation. When the silo is packed slowly, a very decided fermentation occurs by which the mass is raised to a high temperature (1400 F. to 160° F.).

This heating is produced by certain species of bacteria which grow readily even at this high temperature. The fermentation uses up the air in the silo to a certain extent and produces a settling of the material which still further excludes air. The first fermentation soon ceases, and afterward only slow changes occur. Certain acid-producing bacteria after a little begin to grow slowly, and in time the silage is rendered somewhat sour by the production of acetic acid.

But the exclusion of air, the close packing, and the small amount of moisture appear to prevent the growth of the common putrefactive bacteria, and the silage remains good for a long time. In other methods of filling the silo, the food is very quickly packed and densely crowded together so as to exclude as much air as possible from the beginning. Under these conditions the lack of moisture and air prevents fermentative action very largely. Only certain acid-producing organisms grow, and these very slowly.

The essential result in either case is that the common putrefactive bacteria are prevented from growing, probably by lack of sufficient oxygen and moisture, and thus the decay is prevented. The closely packed food offers just the same unfavourable condition for the growth of common putrefactive bacteria that we have already seen offered by the hard-pressed cheese, and the bacteria growth is in the same way held in check. Our knowledge of the matter is as yet very slight, but we do know enough to understand that the successful management of a silo is dependent upon the manipulation of bacteria.

The Fertility of The Soil

The farmer's sole duty is to extract food from the soil. This he does either directly by raising crops, or indirectly by raising animals which feed upon the products of the soil. In either case the fertility of the soil is the fundamental factor in his success. This fertility is a gift to him from the bacteria.

Even in the first formation of soil he is in a measure dependent upon bacteria. Soil, as is well known, is produced in large part by the crumbling of the rocks into powder. This crumbling we generally call weathering, and regard it as due to the effect of moisture and cold upon the rocks, together with the oxidizing action of the air. Doubtless this is true, and the weathering action is largely a physical and chemical one. Nevertheless, in this fundamental process of rock disintegration bacterial action plays a part, though perhaps a small one. Some species of bacteria, as we have seen, can live upon very simple foods, finding in free nitrogen and carbonates sufficiently highly complex material for their life. These organisms appear to grow on the bare surface of rocks, assimilating nitrogen from the air, and carbon from some widely diffused carbonates or from the CO, in the air. Their secreted products of an acid nature help to soften the rocks, and thus aid in performing the first step in weathering.

The soil is not, however, all made up of disintegrated rocks. It contains, besides, various ingredients which combine to make it fertile. Among these are various sulphates which form important parts of plant foods. These sulphates appear to be formed, in part, at least, by bacterial agency. The decomposition of proteids gives rise, among other things, to hydrogen sulphide (H,S). This gas, which is of common occurrence in the atmosphere, is oxidized by bacterial growth into sulphuric acid, and this is the basis of part of the soil sulphates. The deposition of iron phosphates and iron silicates is probably also in a measure aided by bacterial action. All of these processes are factors in the formation of soil. Beyond much question the rock disintegration which occurs everywhere in Nature is chiefly the result of physical and chemical changes, but there is reason for believing that the physical and chemical processes are, to a slight extent at least, assisted by bacterial life.

A more important factor of soil fertility is its nitrogen content, without which it is completely barren. The origin of these nitrogen ingredients has been more or less of a puzzle. Fertile soil everywhere contains nitrates and other nitrogen compounds, and in certain parts of the world there are large accumulations of these compounds, like the nitrate beds of Chili. That they have come ultimately from the free atmospheric nitrogen seems certain, and various attempts have been made to explain a method of this nitrogen fixation. It has been suggested that electrical discharges in the air may form nitric acid, which would readily then unite with soil ingredients to form nitrates.

There is little reason, however, for believing this to be a very important factor. But in the soil bacteria we find undoubtedly an efficient agency in this nitrogen fixation. As already seen, the bacteria are able to seize the free atmospheric nitrogen, converting it into nitrites and nitrates. We have also learned that they can act in connection with legumes and some other plants, enabling them to fix atmospheric nitrogen and store it in their roots.

By these two means the nitrogen ingredient in the soil is prevented from becoming exhausted by the processes of dissipation constantly going on. Further, by some such agency must we imagine the original nitrogen soil ingredient to have been derived. Such an organic agency is the only one yet discerned which appears to have been efficient in furnishing virgin soil with its nitrates, and we must therefore look upon bacteria as essential to the original fertility of the soil.

But in another direction still does the farmer depend directly upon bacteria. The most important factor in the fertility of the soil is the part of it called humus. This humus is very complex, and never alike in different soils. It contains nitrogen compounds in abundance, together with sulphates, phosphates, sugar, and many other substances. It is this which makes the garden soil different from sand, or the rich soil different from the sterile soil. If the soil is cultivated year after year, its food ingredients are slowly but surely exhausted. Something is taken from the humus each year, and unless this be replaced the soil ceases to be able to support life. To keep up a constant yield from the soil the farmer understands that he must apply fertilizers more or less constantly.

This application of fertilizers is simply feeding the crops. Some of these fertilizers the farmer purchases, and knows little or nothing as to their origin. The most common method of feeding the crops is, however, by the use of ordinary barnyard manure.

The reason why this material contains plant food we can understand, since it is made of the undigested part of food, together with all the urea and other excretions of animals, and contains, therefore, besides various minerals, all of the nitrogenous waste of animal life. These secretions are not at first fit for plant food. The farmer has learned by experience that such excretions, before they are of any use on his fields, must undergo a process of slow change, which is sometimes called ripening.

Fresh manure is sometimes used on the fields, but it is only made use of by the plants after the ripening process has occurred. Fresh animal excretions are of little or no value as a fertilizer. The farmer, therefore, commonly allows it to remain in heaps for some time, and it undergoes a slow change, which gradually converts it into a condition in which it can be used by plants. This ripening is readily explained by the facts already considered. The fresh animal secretions consist of various highly complex compounds of nitrogen, and the ripening is a process of their decomposition.

The proteids are broken to pieces, and their nitrogen elements reduced to the form of nitrates, leucin, etc., or even to ammonia or free nitrogen. Further, a second process occurs, the process of oxidation of these nitrogen compounds already noticed, and the ammonia and nitrites resulting from the decomposition are built into nitrates. In short, in this ripening manure the processes noticed in the first part of this chapter are taking place, by which the complex nitrogenous bodies are first reduced and then oxidized to form plant food. The ripening of manure is both an analytical and a synthetical process.

By the analysis, proteids and other bodies are broken into very simple compounds, some of them, indeed, being dissipated into the air, but other portions are retained and then oxidized, and these latter become the real fertilizing materials. Through the agency of bacteria the compost heap thus becomes the great source of plant food to the farmer. Into this compost heap he throws garbage, straw, vegetable and animal substances in general, or any organic refuse which may be at hand. The various bacteria seize it all, and cause the decomposition which converts it into plant food again. The rotting of the compost heap is thus a gigantic cultivation of bacteria.

This knowledge of the ripening process is further teaching the farmer how to prevent waste. In the ordinary decomposition of the compost heap not an inconsiderable portion of the nitrogen is lost in the air by dissipation as ammonia or free nitrogen. Even his nitrates may be thus lost by bacterial action. This portion is lost to the farmer completely, and he can only hope to replace it either by purchasing nitrates in the form of commercial fertilizers, or by reclaiming it from the air by the use of the bacterial agencies already noticed. With the knowledge now at his command he is learning to prevent this waste. In the decomposition one large factor of loss is the ammonia, which, being a gas, is readily dissipated into the air.

Knowing this common result of bacterial action, the scientist has told the farmer that, by adding certain common chemicals to his decomposing manure heap, chemicals which will readily unite with ammonia, he may retain most of the nitrogen in this heap in the form of ammonia salts, which, once formed, no longer show a tendency to dissipate into the air.

Ordinary gypsum, or superphosphates, or plaster will readily unite with ammonia, and these added to the manure heap largely counteract the tendency of the nitrogen to waste, thus enabling the farmer to put back into his soil most of the nitrogen which was extracted from it by his crops and then used by his stock. His vegetable crops raise the nitrates into proteids. His animals feed upon the proteids, and perform his work or furnish him with milk. Then his bacteria stock take the excreted or refuse nitrogen, and in his manure heap turn it back again into nitrates ready to begin the circle once more. This might go on almost indefinitely were it not for two facts : the farmer sends nitrogenous material off his farm in the milk or grains or other nitrogenous products which he sells, and the decomposition processes, as we have seen, dissipate some of the nitrogen into the air as free nitrogen.

To meet this emergency and loss the farmer has another method of enriching the soil, again depending upon bacteria. This is the so-called green manuring. Here certain plants which seize nitrogen from the air are cultivated upon the field to be fertilized, and, instead of harvesting a crop, it is ploughed into the soil. Or perhaps the tops may be harvested, the rest being ploughed into the soil.

The vegetable material thus ploughed in lies over a season and enriches the soil. Here the bacteria of the soil come into play in several directions. First, if the crop sowed be a legume, the soil bacteria assist it to seize the nitrogen from the air. The only plants which are of use in this green manuring are those which can, through the agency of bacteria, obtain nitrogen from the air and store it in their roots. Second, after the crop is ploughed into the soil various decomposing bacteria seize upon it, pulling the compounds to pieces.

The carbon is largely dissipated into the air as carbonic dioxide, where the next generation of plants can get hold of it. The minerals and the nitrogen remain in the soil. The nitrogenous portions go through the same series of decomposition and synthetical changes already described, and thus eventually the nitrogen seized from the air by the combined action of the legumes and the bacteria is converted into nitrates, and will serve for food for the next set of plants grown on the same soil. Here is thus a practical method of 'using the nitrogen assimilation powers of bacteria, and reclaiming nitrogen from the air to replace that which has been lost.

Thus it is that the farmer's nitrogen problem of the fertile soil appears to resolve itself into a proper handling of bacteria. These organisms have stocked his soil in the first place. They convert all of his compost heap wastes into simple bodies, some of which are changed into plant foods, while others are at the same time lost.

Lastly, they may be made to reclaim this lost nitrogen, and the farmer, so soon as he has requisite knowledge of these facts, will be able to keep within his control the supply of this important element. The continued fertility of the soil is thus a gift from the bacteria.

Sustainable Agriculture Focus on Issues Facing Farmers and Producers

The issues that are important to farmers across the country include stewardship, profitability, sustainability and the health of their community. And those are the topics to be covered in Kansas in an upcoming agriculture roundup.

The practice of sustainable agriculture is built upon soil fertility and the protection of soil health. For centuries, farmers around the world have been employing these basic techniques to keep their soil productive. Early in colonial American history, George Washington was using cover crops and manure applications on his farm. American farmers, who did not carefully tend their soil, eventually "wore it out." The Westward Expansion partly reflected the lowered productivity of eastern lands and the search for new farmland farther west.

The first publications from the Kansas State Agricultural College appeared in the late 1880's, at a time when Kansas soils had been tilled for less than fifty years. Soil scientists were well aware at that time that the long term sustainability of farming depended upon the use of legumes for fertility and the addition of organic matter for tilth. As Kansas farmers began to report declining fertility levels in the early 1900's, scientists cautioned that farmers must not rely solely upon the natural fertility of the prairie soils.

Between 1900 and 1950, Kansas State researchers guided farmers in the use of legumes and the preservation of soil organic matter. With the advent of inexpensive nitrogen fertilizers after World War II, traditional soil management techniques took a back seat to the use of commercial products that were easy to use and often provided greater economic returns in the short-term.

At the beginning of the twenty-first century, farmers are becoming increasingly conscious of the importance of soil health, water quality, and energy conservation. The rising cost of nitrogen fertilizers has revived interest in nitrogen-fixing legumes. Excess phosphorus levels in surface water indicate a need to emphasize soil conservation and the careful use of manure resources. Traditional soil management practices continue to be vital for the sustainable agriculturist and are regaining an audience with conventional agriculture.

This review of Kansas State University soil publications profiles the research and recommendations of Kansas scientists during the early to mid-twentieth century. Generally, recommended practices such as crop rotations, manure use, cover crops, and other sources of fertility are considered including the shift to commercial fertilizers in the 1950's. The special consideration of western Kansas soils is treated separately. Limited rainfall in western Kansas affects the research conducted in that part of the state and alters the use of traditional soil health practices.

Soil Health

Within every aspect of Kansas agriculture, healthy soil is a key element. Its structure and fertility provide the basis for all crop and livestock production. Throughout the historical publications of Kansas State University from the late 1800's to the mid-1900's, researchers and educators have been concerned with the protection of this vital resource. In their 1918 publication, Soil Fertility, L.E. Call and R. I Throckmorton caution the Kansas farmer. "The soil is the most important source of wealth in an agricultural state. If it is maintained in a high state of productivity, by wise systems of soil management, the people prosper. If its fertility is wasted through careless methods of farming, both the farmer and the state suffer".

Forty years later, in 1956, Orville Bidwell echoes this same message with its promise of a precarious wealth. "Unlike most other resources, soil is inexhaustible if properly managed." Bidwell recounts the variety of Kansas soils each with a different waterholding capacity, permeability, response to fertilizers, and susceptibility to erosion.

Decline of Kansas Soils

Throughout many of the earliest publications, the authors are clearly concerned about the declining condition of Kansas soils. By the early 1900's, much of the rich prairie soil in eastern Kansas had been farmed for 50 years. A 1903 publication from the veterinary department of the Kansas State Experiment Station states, "The fertility of the soil of the Middle states and the West is being rapidly diminished and if means are not taken to prevent it, the time is not far distant when it will be as necessary to apply artificial fertilizers to the soil as it is now in the East".

A few years later, chemists at the agricultural experiment station raised the same concern. "In the early history of Kansas no attention was paid to the composition of its soils except to boast of their inexhaustible fertility. The voice of the chemist has been lifted constantly, warning the people that this idea of possession of a fertility that is practically limitless is a delusion that can lead only to squandering of our natural resources, and to leaving posterity handicapped in the struggle for existence. Today he is seeing his warnings justified. People in many localities of the eastern part of the state are making inquiry concerning chemical analysis of their soils with reference to learning what fertilizers should be applied and to what crops their soils are best adapted".

By 1918, L.E. Call and R. I. Throckmorton attempted to put dollar figures to the losses in soil fertility. They estimated that the plant food removed from Kansas soils by wheat crops over the previous fifty-five years equalled a value of more than seven hundred million dollars. Even the wheat straw, which was regularly burned or wasted, had a value in plant nutrients of more than twelve million dollars. The majority of the wheat products were both milled and eaten outside the state which Call and Throckmorton equated with the export of soil fertility.

Call and Throckmorton credited the declining productivity of Kansas soils to five factors: depletion of soil organic matter, failure to grow enough acres of leguminous crops for nitrogen fixation, depletion of mineral nutrients, the lack of proper crop rotations, and the erosion of fertile topsoil. These five factors are the subject of nearly every soils publication prior to the advent of inexpensive nitrogen fertilizers in the mid-1900's. They continue to be the basis of soil health for every farming system regardless of the use of commercial fertilizers.

Soil Organic Matter

The early Kansas State publications recognized the importance of soil organic matter to soil health. Replenishment of soil organic matter is the most basic step in addressing a number of other issues. According to Call and Throckmorton, organic matter holds the soil's store of nitrogen, provides good tilth, holds moisture, and provides food for the bacteria that make nutrients available to plants.

Soil studies in western Kansas spanning over 30 years of crop production give mention to the importance of organic matter. Progress reports in 1943 and 1957 indicate that as the organic carbon content of the soil has decreased, more power is needed for tillage, water intake is decreased, seedling emergence and root growth are hindered. The 1943 report also mentions that the presence of coarse, fibrous organic matter helped reduce wind erosion. Numerous publications cite frequent plowing and intensive cultivation as culprits in the loss of organic matter. Row crops depleted the soil more quickly than small grain crops. Without a regular practice of restoring organic matter, the levels of carbon in the soil would drop, threatening fertility and soil structure.

The 1918 publication, Soil Fertility, promotes barnyard manure applied to the soil as a primary source of organic matter. However, when sufficient manure is not available, the farmer should grow a crop to plow under to supply organic matter. These green manure crops may be legumes such as alfalfa, cowpeas, soybeans, clover, or sweet clover. Legumes would capture atmospheric nitrogen and fix it in the soil in addition to supplying organic matter. Non-legume crops such as rye, buckwheat, sorghum, and turnips were also cited as possible sources for organic matter.

Although most of the methods for increasing organic matter focus on the addition of green manure crops or barnyard manure, the conservation of all organic matter is addressed. Throckmorton and Call state that most wheat straw is either burned or destroyed after threshing. They recommend its use as feed and bedding for livestock. Eventually the soiled bedding and any manure would be returned to the soil. They also suggest use of the straw as a surface mulch on wheat during the winter at a rate of 1 - 1.5 tons per acre.

In 1962, farmers were again cautioned not to waste the organic matter provided by straw or stubble. In an effort to control weeds and increase yields, burning of wheat stubble had become a common practice. A series of studies in western Kansas produced data showing that stubble burning did not increase yields for subsequent crops. Not only did it present increased potential for wind erosion, it also decreased the soil's ability to absorb water.

Nitrogen and Organic Matter

The supply of nitrogen in the soil was closely linked to the organic matter content. The practice of returning organic matter to the soil in the form of leguminous plants or animal manures was also the method for supplying nitrogen prior to the use of commercial nitrogen fertilizer. Legume crops, grown in rotation with other crops, could be worked back into the soil as a green manure crops or the legume hay crop was fed to livestock and their manure was returned to the soil.

Off-farm sources of nitrogen in the early 1900's included waste materials from the packing houses and inorganic compounds such as saltpeter, which was mined in Chili, and manufactured compounds of ammonium sulphate or calcium cyanamide. Researchers noted that "the purchase of nitrogenous fertilizers should be limited to the meeting of special requirements of certain conditions or crops...Nitrogen...is the (nutrient) most cheaply restored, since by the cooperation of clovers, alfalfa, peas, beans, and other legumes with bacteria that grow upon their roots the abundant nitrogen of the air in the pores of the soil is brought into organic combination. This means of adding nitrogen to a soil must never be lost to view...".

Through the first half of the twentieth century, researchers and farmers sought to understand the best methods for capturing nitrogen with legumes especially when those crops were being used for other purposes on the farm.

In 1918, researchers were concerned that the nitrogen found within the crop roots and stubble might not be significant compared with the amount removed in a hay crop. They hypothesized that a certain amount of leaf loss during the haying process might be returning some nitrogen to the soil. Still they cautioned that the best practice was to feed the hay on the farm and return all manure to the soil.

The 1939 publication on fertility studies at the Manhattan experiment station beginning in 1910 gave evidence that soil nitrogen could be increased even when the hay was removed. The average increase in the 5-9 year old hay plots was twice that for the 1-4 year plots. The researchers also found that the residual effects of the "nitrogen accumulating capacity" remained for eight to nine years following the breaking of the alfalfa sod that had been in alfalfa for more than two years.

But nitrogen was not the only concern when hay was taken off the farm. "If alfalfa is sold off the land it is one of the most soil-exhausting crops raised, while if it is fed on the farm, and the manure produced applied to the land, it is a conserver of fertility. The same argument applies to clover". Potassium, phosphorus, calcium and other mineral elements are taken up by the plants and must be cycled back to the soil as green manure or barnyard manure. When hay is sold off the farm, the phosphorus and potassium are more rapidly depleted than with a continuous grain crop.

Regardless of the use of green manure crops, any farm with livestock had a source of organic matter and nutrients in animal manure if it was used wisely. "If the livestock farmer properly saves and utilizes his manure he can maintain his soil in a high state of productivity, but the livestockman who feeds his cattle in woodlots along the banks of streams, and so wastes his manure, usually depletes the fertility of his soil more rapidly than the man producing grain only. It was important to manage manure so that nutrients were not lost before they were cycled back to the soil. The seepage of liquid waste or urine, leaching of manure by rain and runoff water, and the "decay" of manure solids and the subsequent loss of nitrogen are all a result of poor handling techniques. Throckmorton and Call recommend that the most practical method of manure handling would be to feed the stock on the cultivated fields so that the manure is scattered by the animals and the nutrients are retained by the soil. In any case, the manure should be returned to the soil as soon as possible and long term, open storage of six months or more should be avoided.

Manure could also be used sparingly as a top dressing on corn, kafir, or winter wheat where it would act as mulch to retain moisture. If there was not a large supply of manure, it was deemed better to apply the available supply lightly to more acres rather than a heavy application on just a few acres.

As a percentage of volume, the nutrients present in manure are small. "The figures for the fertilizing constituents are always low, but they are present in readily available form and the accompanying organic matter has itself a highly beneficial effect on the land. Even with these low percentages the total amount of plant food in the manure produced on a farm reaches very significant quantities".

Phosphorus

Fertilizers were commonly used to provide phosphorus and other minerals long before nitrogen fertilizers were in general use. By 1918, the soils in eastern and southeast Kansas had such low stores of phosphorus that it was profitable to purchase phosphorus fertilizers. Grain farms lost phosphorus the most rapidly when it was exported off the farm with the grain. In order to retain the mineral, the farmer had to feed the grain to livestock and apply the manure to the croplands or else import a supply of phosphorus.

Alfalfa and other deep-rooted crops could be used to collect phosphorus from the subsoil. The nutrients then needed to be cycled back into the upper levels of the soil as either a green manure crop or manure from animals fed on the hay..

Commercial sources of phosphorus included bone, basic slag, rock phosphate, and apatite. Bones were a valuable product rich in phosphorus and nitrogen. They could be ground raw but they were more commonly steamed before grinding. This processing made grinding easier, concentrated the phosphorus slightly and increased the availability to plants.

Basic slag was a by-product from a particular method of iron refinement. Minerals removed from certain iron ore contained high amounts of phosphate, which could be ground for use as a fertilizer. This was not a common product in America, being chiefly available in Europe.

Primary sources of rock phosphate were found in Florida, South Carolina, and Tennessee during the early part of the twentieth century. Although it was most often used to produce superphosphate, it could be finely ground and used raw. Superphosphate was produced by treating the raw phosphate with sulphuric acid and was in use to some extent throughout the twentieth century.

Although superphosphate was more readily available to the plant, Kansas State researchers advised that the raw phosphate was usually a wiser choice. "In the application of phosphate fertilizers the farmer naturally expects and desires immediate results, which are secured by the use of superphosphate, but at the same time if larger quantities of phosphates can be applied in other less soluble and available forms at the same expenditure, the ultimate value of the investment may be much greater, as the phosphorus will remain in the soil and be rendered available by slow natural processes.

There was also indicated a positive relationship between the bacteria found in organic matter and the availability of phosphorus from raw phosphate. "The cheapest source of phosphorus is ground rock phosphate, and in this form it will be available for the use of crops if the soil is well supplied with organic matter from farm manures and legumes. Apatite, a crystal found in granite, was abundantly available in Canada. It needed to be converted to superphosphate to be used as a fertilizer.

Potassium

Potassium occurs in the mineral or rock portion of the soil. Plant roots are able to access potassium directly from tiny soil particles, chiefly silt and clay. Potassium is also present in decaying crop residues and organic matter, indicating the need, once again, for returning plant materials to the soil. A traditional source of additional potassium was hardwood ashes. Muriate of potash, a processed form of potassium in wide use today, was also available early in the twentieth century but due to the chlorine content of the compound, its use was restricted with certain crops.

Calcium (Lime)

Calcium, while an essential element for plant growth, was usually not deficient in Kansas soils. However, its application in the form of lime, was a common soil amendment. "(Calcium) is generally present in all cultivated soils in sufficient quantity to supply fully the need of the plant. Yet even where this is the case, the soil may be greatly in need of liming. Lime is used, therefore, as a soil amendment not so much for its effect directly on the plant as for its effect on the soil, which indirectly affects the plant. Soils that are low in lime are said to be sour". The acidity of these soils could be neutralized by the application of calcium compounds.

A 1918 publication recommended two tons of ground limestone per acre followed by one or two tons every five to six years thereafter. The limestone should be worked into the bare ground six months to a year prior to seeding alfalfa or clovers. The effects of the limestone are slow and gradual but long lasting. The bacteria that live on the roots of alfalfa, sweet clover, and clover thrive in more alkaline soils so these crops respond well to liming. Crops considered less sensitive to acid soils were corn, wheat, timothy, and oats.

The change in soil pH improved the availability of phosphorus and potassium as well as improving the texture of the soil. "It is a well-known fact that soils well stocked with a supply of lime will be more productive under the same conditions of plant-food content than soils not so stocked...While soils use very small amounts of lime as compared with phosphorus and potassium, yet the presence of a relatively large supply of lime insures crop production. In the words of Hilgard, 'A lime country is a rich country'".

Soil Bacteria

Although the exact relationship did not seem to be clear, researchers in the early part of the twentieth century were writing about a correlation between humus or organic matter, bacteria, and fertility. A series of studies beginning in the 1890's showed crop yields in "direct proportion" to the bacterial content of the fields. In this same publication, the authors speculate that fertility depends to a large extent on bacterial activity and that by manipulating the bacteria of the soil, one might avoid the need for artificial fertilization. They proposed additional experiments to find ways to increase soil bacteria.

A few years later, Walter King and Charles Doryland report on their studies of soil bacteria with this same basic assumption. Assuming that bacterial activity indicated increased soil fertility, they set about collecting samples from various soil depths on plots representing a variety of tillage practices. They admitted tremendous variability in their data, which was collected over a period of only three months from

March to June 1908. Nevertheless, they concluded that deep plowing, conveniently the tillage practice most commonly used at that time, increased bacteria levels and bacterial activity and decreased denitrification. They noted that bacterial activity increased with the temperature of the soil and decreased when the soil became saturated with moisture. Different species were predominate at different times and activity seemed to rise and fall with a regularity independent of moisture and temperature.

Analysis of the Soil

As farmers began to experience decreasing yields on exhausted soil, they began to show interest in chemical analysis of their soil. Inexpensive, reliable soil tests were not yet available in the early part of the century. Throckmorton and Call noted that although they could determine the nutrient needs of a crop and the chemical analysis of the soil, their current methods did not tell them what nutrients were available to the crop. They felt that available nutrients "fluctuate greatly" depending upon the total nutrients in the soil, the organic matter content, the weather, cultivation methods and the current crop. Consequently, a chemical analysis would only give a farmer a general outline. Probably the greatest deterrent to using a chemical analysis was the expense which made it impractical for individual farmers.

In 1909, the Extension Station council authorized the Chemistry Department to collect and analyze typical soils from across Kansas. This body of work did give farmers an idea of the basic properties of their soil and requirements of common crops even if it could not provide specific answers regarding fertility.

In 1910, the Chemistry Department cautioned that chemical analysis of the soil was not a reliable indicator regarding crop needs. They recommended that the farmer should test his soil by raising small crop plots that had been "fractionally fertilized" in order to determine the optimal rates. The complexities of variable rate plots involving a significant amount of land and more than one growing season probably doomed this testing method for on-farm use.

More practical advice involved a farmer's observational skills. "Chemical and physical investigation of soils...must be supplemented or...replaced by observations upon the natural growth of trees, shrubs, grasses or weeds upon the soil, and by experiments in the production of plants or crops upon it. Let organic nature answer the question, What is this soil good for?

Observations concerning the natural plant growth upon a soil have always been used by practical men in judging of its value...This means of gaining an insight into soil values is one that, while used from time immemorial, is worthy of more extended study and application

Commercial Fertilizers

Throughout the Kansas State publications from the first half of the twentieth century, farmers are cautioned about reliance on chemical fertilizers. "It should not be forgotten...that barnyard manure, because of its content of organic matter in a state of decay, is superior to chemical fertilizers containing equal amounts of potassium, phosphorus, and nitrogen compounds".

In 1918, Throckmorton and Call warned that although commercial fertilizers are more concentrated, they do not supply organic matter, which is "absolutely necessary to supply plant food, to preserve good tilth, and to retain water in the soil." Because commercial fertilizers do not supply organic matter, "they can't be expected to replace manure in soil improvement, but should be used, where they can be used profitably, in addition to barnyard manure and other forms of organic matter". This caution continued for more than thirty years while Kansas farmers were advised to use legumes and manures as nitrogen sources and pay careful attention to nutrient cycling on their farms. In 1956, the Kansas State

Agricultural College publication, "Legumes vs. Commercial Fertilizer" documented a major change on Kansas farms. The authors reported on a study of the economics of Kansas cropping systems. One objective was to understand why farmers were planting fewer legume acres than were recommended. "Data indicate considerable advantage to certain legume rotations...Even though this is more profitable, farmers probably grow fewer legumes because they need income quickly".

Once a farmer began using commercial fertilizers, the soil would be slow to return to a more natural cycle of fertility that did not include commercial fertilization. In 1918, farmers were already asking whether commercial fertilizers "impoverish the soil." There were reports that farmers who stopped using commercial fertilizers experienced reduced crop yields. Throckmorton and Call explained that these fertilizers "cannot in themselves be expected to maintain the fertility of the soil. They should, therefore, be used only when a good rotation of crops is practiced, and when organic matter is supplied systematically".

Crop Rotations

In the 1910 bulletin, "Fertilizers and Their Use," the authors speculate that there is something other than the chemical analysis of the soil, which affects plant growth. They expected that rotations of crops may have an impact on "soil conditions". Without fully understanding the dynamics of rotations, farmers and researchers already considered them an essential part of good soil management.

By 1935, researchers were refining their view of the use of rotations. "Rotation of crops should not be loosely recommended without stating specifically what the rotation should be, or having in mind the wide differences existing between possible rotations". Citing studies of soil fertility under various cropping patterns and soil treatments over a period of twenty years, Throckmorton and Duly emphasize that not all rotations can be used interchangeably.

Some rotations are less effective than others at maintaining soil fertility as well as at providing economic returns. In some instances, continuous cropping of hay or small grains for a few years was better for the soil or the pocketbook than rotations, which included row crops. Soil quality concerns aside, the prices of individual crops and their cost of production weigh heavily in determining the most profitable rotations. As prices fluctuate, the rotations providing the greatest economic return also vary.

During the latter half of the 1930's, an analysis of these continuing soil fertility studies in Manhattan looked at nitrogen and organic carbon levels. One conclusion was that within the cropping patterns studied, "the larger the percentage of the crop cycle occupied by biennial or perennial legumes or sod crops, the higher will be this level (of nitrogen and carbon)." The use of manure, fertilizers, and lime, if they stimulated crop growth, particularly of a legume, would maintain higher levels of nitrogen and carbon.

Wrong Turns

Besides repeated references to the importance of organic matter in retaining soil moisture, researchers have addressed other means of moisture conservation. A very early bulletin from 1899 examined the possibility that fertilizers themselves might slow water evaporation from the soil. Various treatments were tried on outdoor plots and on small pots within the laboratory over a number of years. In every case, there was no difference in treatment.

Perhaps the most intriguing study was sponsored by the E.I. DuPont de Nomours Powder Company from 1911 to 1913. The dynamite industry had been heavily promoting the use of their product for improvement of all types of soils. The study included plots on heavy clay soils at the Fort Hays and Manhattan experiment stations as well as on numerous farms across the state. On a field eighty rods long and eight rods wide, thirty-inch holes were dug fifteen feet apart in rows sixteen feet apart. One half stick of dynamite was placed in each hole and exploded.

Data showed no significant differences in crop yields, soil moisture, nitrate levels, or bacterial activity on the dynamited soil. Unfortunately, the physical characteristics of the soil were considerably diminished. The explosion forced soil at the centre of the charge into the surrounding pore spaces producing a cavity surrounded by a hard, compact mass. These "jugs" would fill with water during rainstorms and hold it until evaporated.

The cost was prohibitive with the dynamite expense alone at $12.20/acre. Labour was an additional $5.00/acre. The researchers concluded, "In no instance was there improvement sufficient to pay the expense of dynamiting".

Erosion Control

Erosion was recognized as one of the factors resulting in soil depletion. Throckmorton and Call felt erosion could be prevented by deep plowing, adding organic matter and by "working the ground at right angles to the slope of the land". Although plowing would later be seen as a culprit in erosion, it may have offered an advantage over shallow disking by initially creating a rough surface more resistant to wind. The merits of organic matter in stabilizing the soil were universally touted throughout early publications. However, in 1957, researchers in western Kansas reported findings indicating that increases in the soil's organic carbon content did not necessarily mean less susceptibility to wind erosion. Heavy soils could be more vulnerable with the addition of well-decomposed organic matter. Undecomposed crop residues or other organic matter were beneficial in slowing the effects of the wind.

Farmers could create another tool to control wind erosion by alternating strips of crops. Researchers in the 50's set out to determine the ideal width of these strips for maximum protection. They considered the quantity of crop residue and soil roughness that would be produced during years of low rainfall, high wind, and low crop yields. Full wind protection during such years required strips so narrow that they would be impractical to farm. When combined with other erosion control methods such as high residue management, the field strips could be increased in size to an acceptable level and still provide a high degree of protection.

A 1962 bulletin on farming systems in western Kansas notes that contour farming resulted in increased yields for a wheat, wheat, sorghum, and barley rotation at Fort Hays. These researchers also experimented with dikes around very level fields to capture moisture and found some yield advantage.

Western Kansas Cropping Systems

Kansas State research bulletins and publications documenting the work in western Kansas over the first half of the 1900's reflect a very different type of cropping system from that used in the eastern parts of the state. Dryland farming in areas of low rainfall and high winds presents special challenges for soil health. This flat, arid region in the western half of the state supports a fragile wealth that requires careful consideration to maintain soil resources.

Soil studies in western Kansas date back to the very earliest years of the twentieth century. A continuous study of organic carbon and nitrogen in soils at Ft. Hays Experiment Station lasted for more than thirty years with follow-up research continuing for at least another fifteen years. The western Kansas studies covered a number of topics including the use of fallow periods, changes in organic carbon and nitrogen, tillage methods, and crop rotations. In nearly every instance, the principle concerns were the conservation of soil moisture and fertility.

Fallow

Fallow is the "practice of keeping land free of all vegetation throughout one season for the purpose of storing moisture for a crop the following year. Where rainfall is too low to support yearly crop production, fallow can be an important part of the cropping system.

Using fallow to store soil moisture results in production stability by decreasing the number of crop failures. Adherence to the fallowing pattern in high rainfall years is important since a portion of that moisture is stored for following dry years.

In 1962, Ft. Hays researchers reported that milo yields after fallow were twice the yields of milo crops following milo. Although this was the greatest percentage increase for any crop studied, all crops saw increases (1962, Investigations of Cropping Systems, Tillage Methods, and Cultural Practices for Dryland Farming).

A 1941 publication states that corn, oats, and barley grown on fallowed ground show a marked increase in quality even if the yield from one year does not equal two crops. Fallow in a rotation with forage crops "allows farms without pasture to reintegrate livestock and supplement the carrying capacity of those farms with pasture".

The successful use of fallow is related to soil type with its greatest value seen on heavier soils that have an increased moisture storage capacity. Light soils without a heavy subsoil, shallow soils, and hilly topography are not likely to provide an economic advantage with fallowing since moisture cannot be held.

Soil management techniques for fallow rotations must always be directed toward capturing moisture, preventing evaporative losses, and timely destruction of weeds. The average rainfall in western Kansas may be adequate to produce a crop however the moisture losses from evaporation, runoff, and weed pressure rob a large portion of the water before it can be utilized. Throckmorton and Myers hypothesize that the farmer has little control over evaporative losses which can be 60-75% of total precipitation. "Fallow is extravagant in so far as storage of total precipitation is concerned, but it is essential as a means of stabilizing production through having sufficient moisture in the soil at seeding time to justify the seeding of a crop".

Tillage is a key to the successful management of runoff and weed growth. "A good summer fallow is one in which the soil is free of all growing plants throughout the fallow period and has a rough open surface which will permit a ready and rapid penetration of moisture." If possible the stubble of the preceding crop should be left standing during the winter and spring to capture snow and prevent wind erosion. Thereafter, cultivation should be used when weeds are still small and the soil will form clods to create a rough surface.

To further protect the soil from wind erosion, fallow strips can be alternated with crop strips following field contours. The fallow that is poorly managed is doubly exposed to the wind. Proper management is an effective means of checking erosion.

In 1962, Luebs adds that shallow cultivation using a one-way disk plow or a subsurface tillage tool is just as effective as a plow or a lister for destroying weeds. This type of tillage leaves plant residues on the surface of the soil to slow the wind action.

Organic Matter Content of Western Kansas Soils

General soil management techniques would indicate that increasing the organic matter content of the soil would be one tool to build its water holding capacity. Long term studies of organic carbon and nitrogen levels in soils at Colby, Hays, and Garden City provide an interesting look at organic carbon levels.

All cropping systems examined in the studies were resulting in decreases in both soil carbon and nitrogen levels. The cropping systems were generally depending upon native fertility of the soil or some applications of manure for crop production. Continuous small grain production or small grains in rotation with a fallow period showed the least destruction of soil organic matter levels. The practice of fallowing in itself decreased organic matter levels since nothing was allowed to grow on the soil but when fallow was used in a rotation with small grain crops, the losses were decreased.

The use of green manure crops to build organic matter might be considered in a higher rainfall area. However, as early as 1918, researchers warned western Kansas farmers that green manure crops would use too much moisture prior to the main crop. They recommended finding some other source of organic matter.

Sampling for the long-term studies began in 1916 and in each successive report, researchers found that applications of animal manure and/or straw slowed the loss of organic carbon and nitrogen or provided a small increase. The relationship with regard to crop yields was less positive.

In 1943, researchers reported that manure applications increased yields only under "certain conditions" and added that "perhaps (manure's) advantage will become more evident in the future. In 1957 manure and straw applications showed no benefit to yield data. Citing the maintenance of nitrogen and organic matter content, researchers concluded that "these results indicate, even at the present time, that all manure should be conserved and applied to the land.

In 1962, researchers stated the manure was of "negligible" value for wheat and sorghum in a fallow-wheat-sorghum rotation. In 1965, after more than forty years of soil studies at the Ft. Hays Experiment Field, the researchers acknowledge that green manures and animal manure applications lower nitrogen and organic carbon losses. But they maintained that their use to maintain soil productivity "in this area is not practical" citing the need for 25-30 tons of manure per acre every three years to maintain nitrogen and carbon levels.

The data regarding crop yields and carbon levels in these studies did not vary significantly over the course of forty years. However the conclusions regarding "practicality" are considerably different. It may be that as farming practices changed and manure was less accessible on the average farm, its use was, indeed, less practical. As livestock concentration has increased in some parts of western Kansas at the end of the twentieth century, we are once again revisiting the practicality of manure applied to cropland. Although the economics may be driven by a need to dispose of excess manure, wise use indicates the "practicality" of using those manures to build fertility and organic matter levels.

Fertility of Western Kansas Soils

"From a practical standpoint, the question of the use of nitrogen fertilizers in western Kansas presents a number of important problems." Researchers in 1943 could see that frequent crop failures and low yields were due to limiting factors other than fertility - principally moisture. They also felt that the use of summer fallow "will reduce the need (for fertilizers) due to the accumulation of nitrate nitrogen". For years they examined nitrogen changes in the soils and considered the best course for maintaining fertility at a level that permitted crop production with an economic return.

The first sixteen years of the long term soil studies at Hays, Colby, and Garden City showed nitrogen losses from 1916-1938 that were nearly equal to the nitrogen removed by the crops. The losses immediately following sod breaking were greater than the losses later in the study. For a time, there was hope that this trend indicated a possible equilibrium for nitrogen levels.

Nitrogen lost through crop removal was being somewhat offset by the deposition of nitrogen in rain and snow during the fallow period. However, researchers estimated this amount to be only three to eight pounds per acre per year with an average deposition of 3.44 pounds per acre each year. They also speculated that there might be some fixation of nitrogen by free-living bacteria (Azotobacter and Clostridium) although they had not been able to verify this was true in any field experiments.

Crop yields in these early studies fluctuated so much depending on rainfall patterns, no trend in decreasing yields due to fertility could be tracked. It is most likely that farmers were still mining the native fertility of the soil but at a much slower rate than the farmers in the eastern half of the state due to the difference in rainfall. Considering the fragility of western Kansas cropping systems, farmers in that area need to be able to adapt their cropping patterns to fit the current conditions. "The successful dryland farmer in this area must, as far as possible, be flexible in choosing cropping sequences, tillage methods, and cultural practices". He or she must consider the weather, soil conditions, preceding crops, residue, weed populations, soil moisture, tilth, and the removal of nutrients.

Looking to the Future of Kansas Soils

Although healthy soil is the basis for the rich diversity of agriculture in Kansas, a host of barriers, both economic and social, prevent us from protecting and building soil quality. Short term leases of rented farm ground discourage the use of practices that only see an economic return after multiple years. Inexpensive commercial fertilizers have precluded the need to monitor nutrient cycling on the farm. New interest in protecting water quality, conserving water use, managing excess livestock wastes, and developing farming systems that reduce the use of commercial fertilizers and pesticides may refocus attention on some of the same questions that were addressed during the first half of the twentieth century. Farmers may gain understanding from the early research of Kansas State University and then begin to ask the questions that will lead us into the twenty-first century with renewed interest in our soil.

Role of Bacteria in Coffee Plantation Ecology

To date we have written a dozen or more articles on the pro active role of microorganisms in shaping the destiny of the Coffee Mountain. However, off late we have received hundreds of e. mail requests from all over the globe, asking us to elaborate on the type of microorganisms carrying out these functions. Armed With a back ground of Microbiology and Horticulture, we have made an honest attempt to simplify the role of BACTERIA in coffee plantation ecology.

All living organisms are classified as either PROKARYOTIC (PRIMITIVE) or EUKARYOTIC (Higher forms of life) based on their cellular structure. Bacteria and blue green algae are grouped under prokaryotes and all other organisms are eukaryotes. Bacteria are unicellular microscopic or single celled organisms widely distributed in nature. The greatest benefit in studying bacteria is that it throws light on the evolution of simple cellular systems and the higher forms of life.

Microorganisms are categorized into six distinct groups.

- Bacteria
- Fungi
- Actinomycetes
- Algae
- Protozoa
- Viruses.

The primary purpose of writing this article is to help coffee farmers worldwide in understanding the basic premise on which BACTERIA operate and the ways and means of carefully exploiting their potential to maintain a perfect ecological balance within the coffee farm.

The coffee habitat provides a fertile ground generating a host of both macro and micro organisms. The bacteria are the most dominant group of microorganisms in coffee soils. The microscopic analysis of coffee soils shows that there is plenty of room at the bottom for the proliferation of different types of microorganisms because of the rich humus and organic matter content. Among the different groups, bacteria are capable of harvesting atmospheric nitrogen, solubilisation of rock phosphate and in the transformations of various substrates resulting in periodic supply of available nutrients for plant growth and development.

However, coffee farmers need to understand that the types of bacteria and their numbers are governed by the soil type and cultivation practices like addition of chemical manures, pesticides, poisons, type of tillage etc. In turn the activity of bacteria is influenced by the availability of nutrients, both in organic and inorganic forms. Their numbers are very high in coffee soils with large number of trees compared to open meadows. This is due to the shading nature as well as greater root density and the abundant availability of soil organic matter. Due to the high organic matter content of coffee soils, bacteria decompose the organic matter and in the process acquire energy.

Bacterial cells are so very small that when you think of these fastidious and ubiquitous microbes, you need to think small. They are measured in microns and the equivalent of one micron is: one by thousandth of a millimetre. Hence one needs a fairly high powered microscope to observe these minute wonders. However, they have a remarkable advantage because they have their strength in numbers. The population of bacterial cells in soils is always great. Due to their rapid growth and short generation time they can quickly act on various organic materials. In harsh environments lacking oxygen the bacteria alone are responsible for almost all the biological and chemical changes. Because of their very small size bacteria have a very high ratio of surface area to volume.

Also, since bacteria are single celled microorganisms they absorb their nutrients through their cell membrane, there by exhibiting very high metabolic rates.

The earliest inhabitants of Planet Earth have undoubtedly been the microorganisms. From primitive prokaryotic unicellular microorganisms, evolved the higher forms of life. Microorganisms have thus been the earliest participants in shaping various life processes. Microorganisms have been largely responsible in changing the primordial atmosphere resulting in the formation of gaseous oxygen needed for plant growth. Fifteen million years of evolution has shaped the coffee forest and today their future is in our hands. Fundamentally, it has been an evolution of skills.

The evolutionary ladder points out to the pivotal role played by microorganisms in adapting to harsh environmental conditions, formation of tripartite bonds, break down of complex polysaccharides into simpler molecules and in the process providing the energy needs of the biotic community. It is in this context that this article throws light on the role of microorganisms, starting with BACTERIA in transforming the coffee landscape into an evergreen rich forest. Among the different microorganisms, Bacteria are known to play a vital role in the distribution and supply of energy needs of the entire coffee mountain. Decomposition of almost all insoluble salts is mediated by one or the other group of bacterial communities.

Distribution and Functions of Bacteria

Bacteria are widely distributed along the length and breadth of the coffee mountain. In short they are found almost every where. Bacteria are single celled organisms and in spite of their simplicity are highly efficient. Their numbers decline with depth of soil.

A majority of the coffee farmers are unaware that the great majority of bacteria are beneficial and absolutely necessary to convert farm wastes, organic debris and other by products into energy rich compounds needed for plant growth and development. Plants and animals depend on the fertility status of the soil and this in turn is dependent on the activity of soil microorganisms. Plants cannot directly utilize organic compounds such as fatty acids, lipids, carbohydrates and proteins. Microorganisms are a vital link in the mineralization of organic constituents and provide nutrients in the available form for plant growth and development. Bacterial cells can withstand long periods of drought due to the protective cover around the cell wall known as CAPSULE.

The capsule is a slimy or a gelatinous material and encloses either one cell or a group of cells. At times the bacteria make use of the polysaccharides present in the capsule as a source of reserve food material. The capsule enables the bacteria to avoid predation by larger soil microbes and infection from viral strains. In addition to protection, capsules also play an important role in the attachment of bacterial cells to plant or rock surfaces and in the formation of biofilms.

Bacteria are morphologically grouped into three types. Cell structure is a key element in the characterization of bacteria.

1. Cylindrical or Rod shaped commonly referred to as Bacilli. They are the most numerous. Bacillus species are known to overcome extreme weather conditions by the formation of endospores that function as part of the normal life cycle of the bacterium. These endospores are resistant to long periods of drought and desiccation. With the on set of favorable conditions the spore germinates and a new bacterial cell grows.
2. Communication between all life is essential in unfolding the various patterns of life. Without the ability to communicate, life, including the simplest single celled organism, could not exist. Bacteria have the capacity to analyze vibrations in the surroundings and accordingly react. Besides shape and size certain rod shaped bacteria have thin hair like appendages on the outer cell wall known as flagella, which can sense the external environment and constantly send out chemical signals to reach out to other communities. Flagella are believed to be organs of locomotion.
3. Spherical or ellipsoidal bacteria are called cocci. Ellipsoidal bacteria occur in pairs and are referred to as STREPTOCOCCI, when in four cells, arranged in a square they are known as TETRADS; when in irregular clusters like a bunch of grapes they are called STAPHYLOCOCCI and when arranged in a cubical form known as SARCINAE.
4. *Spiral or Helicoidal*: Winogradsky a leading soil microbiologist placed Soil bacteria into two broad divisions.

Autochthonous Species

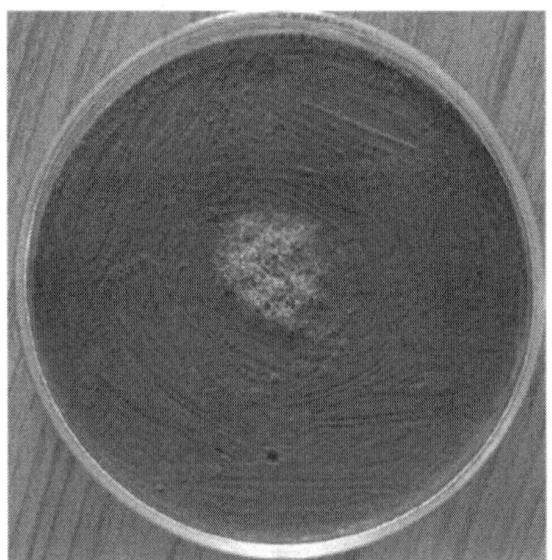

These refer to the indigenous or native species. The population of these bacteria is always uniform and constant in coffee soils because their nutrition is dependent on the native soil organic matter.

They multiply rapidly in the presence of large quantities of biomass, organic matter, humus, and other soil amendments having a low C:N ratio. They are pretty tough and resistant to varied agro climatic conditions. They participate in all biochemical functions of the community. The presence of these bacteria is fairly high and their numbers are constant. The presence or absence of specific nutrients does not change their numbers significantly.

Allochthonous Species or Zymogenous Bacteria or Fermentative: Commonly referred to as the invaders. Their participation in biochemical functions is insignificant.

These bacteria are active fomenters and need nutrients which are quickly exhausted. They are involved in a process in which organic matter is rapidly attacked in successive stages and made available to the plants.

At each stage of decomposition a specific group of organism is involved. The bacterial numbers increase rapidly whenever furnished with the special nutrients (leaf litter, biomass, compost) to which they are adapted. On exhaustion of these nutrients their numbers decrease and return with the addition of nutrients. Hence, this group of bacteria requires an external source of energy for their multiplication and growth.

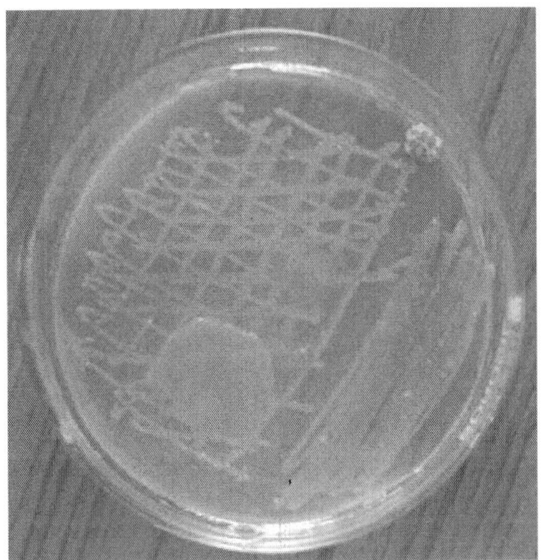

Bacteria in this group include the nitrogen fixers, phosphorus solubilisers, nitrifiers, cellulose hydrolyzing bacteria, sulphur oxidizers, spore forming bacillus and non spore forming pseudomonas.

Environmental Factors

Bacterial numbers, their density, type and composition is governed by the environmental Stimulus.

Aeration

Bacteria are further divided as

Aerobes: Require the presence of oxygen for growth and metabolic activity.

Anaerobes: Bacteria which grow in the absence of oxygen.

Facultative Anaerobes: Develop either in the presence or absence of oxygen.

Aerotolerant Anaerobes: These bacteria grow under both aerobic and anaerobic conditions.

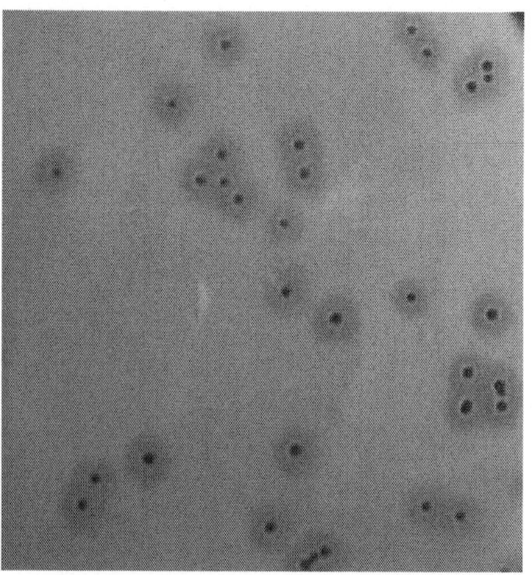

Moisture

Aerobic bacteria are the main stay in coffee soils and the optimum level of moisture content for their activities is in the range of 50 to 75% of the soil's moisture holding capacity. Coffee soils are inherently shaded by tree canopies as well as by the coffee bush. Hence they remain shaded most of the time.

Also, a host of factors result in the availability of moisture throughout the year. For e.g. The south west and the north east monsoon together keep the soil moist for eight months of the year and the remainder months, due to soil conservation practices adopted by the Indian coffee farmer , the moisture is always available for bacterial growth and development. Water makes up a major component of the microbial cell. Hence it is a key component for the functioning of the cell.

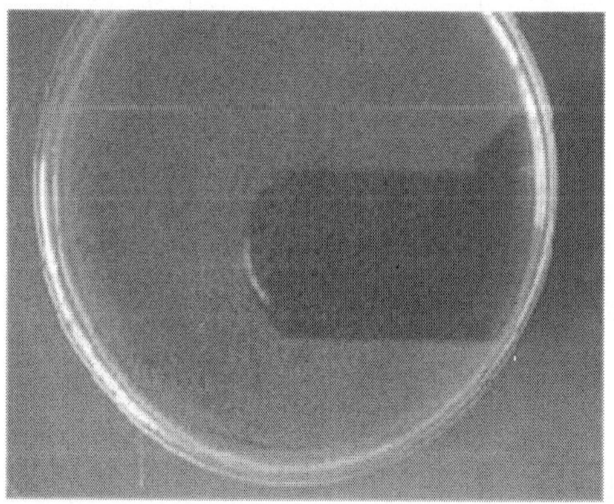

The most common problem encountered in coffee soils is not the lack of moisture but the availability of excess moisture which is detrimental for the growth and multiplication of bacteria. Excess moisture limits the supply of gaseous oxygen resulting in an anaerobic environment. Water logging brings about a decrease in the abundance of bacteria.

Temperature

Bacteria are highly sensitive to temperature fluctuations. Apart from growth and development, temperature plays a vital role in the biochemical processes carried out by the bacterial cell.

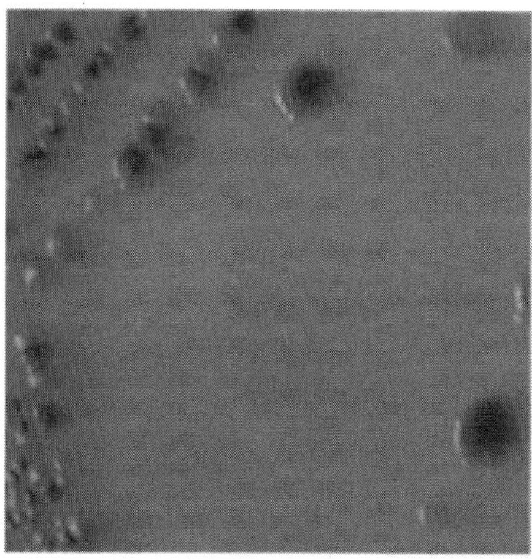

Mesophiles are the ones which grow well in the temperature range of 25 to 35 degree centigrade. These constitute the bulk of the coffee soil bacteria. For most part of the year the temperature profile in the coffee mountain falls in this range. Some scientists have further divided mesophiles as :

Oikophilic; Organisms whose optimum temperature is around 20 degree centigrade. Somatophilic; Organisms whose optimum temperature is about 37 degree centigrade.

Psychrophyles are bacteria that love cold and grow at temperatures below 20 degree centigrade. Thermophiles are temperature loving bacteria and grow best in the temperature range of 45 to 65 degree centigrade. These bacteria are active in compost pits.

Organic Matter

The population of bacteria is directly related to the organic matter content of the soil. Due to periodic leaf shedding and availability of huge quantities of carbonaceous materials on the floor of the coffee forest, the bacterial numbers is the largest. Also the coffee farmers incorporate green manures, compost and biomass from time to time which act as stimulants for the growth and proliferation of bacteria.

Acidity

The optimum pH for the growth of bacteria is NEUTRAL pH. Coffee farmers need to keep the hydrogen ion concentration of their soils close to neutral because in highly alkaline or highly acidic conditions the growth and multiplication of bacteria is inhibited.

In general in heavy rainfall areas receiving 100 inches and more it is advisable to apply lime or dolomite once every two years and in moderate rainfall regions, once every four years. This practice will not only increase the bacterial numbers but will also enable the coffee bush to take up inorganic nutrients in a more efficient way.

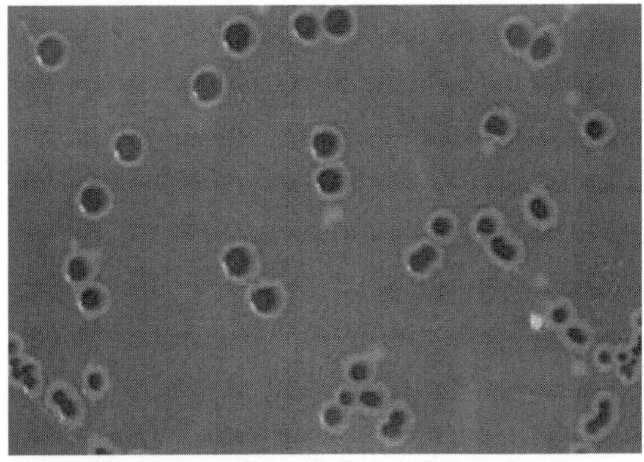

Figure : Acidophilic Bacteria: Bacteria capable of growth in extremely low Ph.

Alkalophilic Bacteria: Bacteria capable of growth in extremely alkaline soils (p H 10.5)

Halophilic Bacteria: Bacteria capable of tolerating high salt concentrations.

Xerophilic Bacteria: Bacteria capable of growth in dry habitats.

Inorganic Nutrients

Application of fertilizers and chemicals greatly affects the bacterial population. Coffee farmer's world wide use ammonium fertilizers as the bulk of fertilizer application. Coffee farmers do not realize that ammonium fertilizers tend to lower the soil pH resulting in acidity due to the microbial oxidation of ammonium to nitric acid.

More than the effect of fertilizer, it is the acidity which suppresses the bacterial population. This problem can be easily overcome by split applications spread out over a two week period. More importantly, the application of fertilizer should be carried out when the soil moisture is optimum. It is a proven fact that small amounts of inorganic fertilizers supply the needs of the bacterial community in the form of inorganic nutrients.

Farm Practices

Farm practices also exert direct and indirect biological effects on the coffee farm. Periodic soil disturbance will affect the bacterial population. Addition of organic manures from time to time and incorporating legumes into the soil with proper carbon nitrogen ratio accelerates the build up of beneficial micro flora. However, if soil hardens up over a period of time, then it will have an adverse effect on the bacterial numbers.

Nutritional Requirements of Bacteria: Macronutrients

Carbohydrarates, Proteins, Lipids, Nucleic Acids. Carbon requirement is the greatest, followed by nitrogen, phosphorus and sulphur. Potassium, sodium, calcium and magnesium are also required in substantial quantities.

Micronutrients

Cobalt, iron, zinc, copper, molybdenum, manganese.

Certain bacteria require specific organic compounds that they are unable to synthesize from simple compounds. Hence they require

GROWTH FACTORS classified into one of the following groups; Amino Acids; Purines & Pyrimidines; Vitamins.

The dominant groups of bacteria are the heterotrophs but a few genera have photo chromatic pigments enabling them to have photoautotrophic nutrition. Two main classes of bacteria are the Heterotrophs or Chemoorganotrophic: Bacteria which require preformed an organic nutrient which serves as a source of energy and carbon.

Autotrophic or Lithotrophic: Bacteria which obtain their energy from sunlight or by the oxidation of inorganic compounds and their carbon by the assimilation of carbon dioxide. Autotrophs are further classified as Photoautotrophs or Photolithotrophs; where energy is derived from sunlight and Chemoautotrophs or Chemolithotrophs which obtain their energy from the oxidation of inorganic materials.

Specialised Bacterial Cells

Endospores: Bacterial communities have developed their own specialized skills to survive the hardships of nature. At times it involves a constant battle where it is not only the survival of the fittest but survival by way of forming alliances with other biotic communities. It involves constant signal exchange between predators and prey, struggles for dominance, defence of territories and many ways to simply survive by the production of spores and endospores which are tolerant to adverse weather conditions.

These structures are resistant to heat, desiccation, high salt concentrations, cold, osmosis and chemicals, compared to the vegetative cells producing them. Endospores are bodies produced within the cells of a considerable number of bacterial species. Sporulation confers protection to the cell whenever the occasion arises. Because of their low rate of metabolism , endospores can survive for a number of years without a source of nutrients.

However, when favorable conditions appear, endospores begin to germinate within a few minutes to form a new vegetative cell.

Most Commonly Encountered Soil Bacteria

Belong to the Following Genera

Aerobacter, Myxobacteria, Bacillus, Pseudomonas, Flavobacterium, Arthrobacter, Achromobacter, Clostridium, Corynebacterium, Mycobacterium, Sarcina, Myxococcus, Archangium, Chondrococcus, Cytophaga, Sporocytophaga, Polyangium.

Figure: *Chemical Composition of the Bacterial Cell on Dry Weight Basis.*

The major constituent is water to the extent of 90 %. The ash content varies. Carbon 45-55% Nitrogen 8-15%. The ash from bacteria may contain Phosphorus 10-50%, Potassium 4-25%, Sodium 10- 35%, Magnesium 0.1-10%, Silica 0.5-7.7%, Calcium 0.3-14%, Chlorine 1-44%, and trace amounts of iron. These variations are due to the different bacterial species on the floor of the coffee mountain.

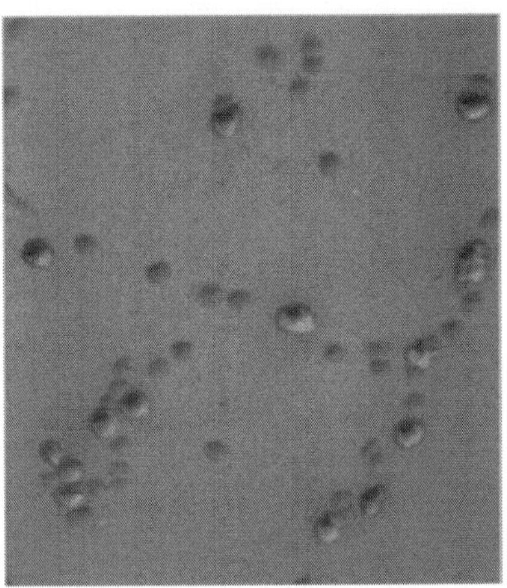

Isolation

We have had the rare opportunity of isolating thousands of species of soil bacteria during our post graduate studies. Our study included soils from various agro climatic regions of Karnataka. Our results had far reaching consequences. We were able to prove that soils with neutral pH harboured the maximum load of nitrogen fixers and phosphate solubilisers. Different soil types had different species of bacteria but most of them were beneficial and acted as important links in various soil transformations.

Conclusion

Shade grown Indian coffee plantations have sustained many generations of farmers because of their resilience in overcoming all odds. The secret behind this success is attributed to the microbial inhabitants. Microorganisms cannot be seen by the naked eye, yet they constitute about one quarter of the biomass-the total weight of living organisms in the world. Animals and plants account for the remainder. In the strict sense, more than 98 % of what we describe as waste inside the farm is valuable food for one or the other group of microorganisms. Bacteria recycle these wastes into power packed energy rich nutrients required for the survival of the coffee bush and its partners. Coffee farmers have an erroneous concept of the role of bacteria in nature. They strongly feel that the majority of bacteria are disease producing. We would like to set the record straight and state that the vast majority of bacteria are not only beneficial but are absolutely essential in building up a healthy coffee farm. Yes, there are a very few bacterial species that are harmful but in a healthy ecosystem they rarely express themselves.

It is very important that the coffee farmer understand the subtle role played by bacteria in the transformation of major elements like nitrogen, sulfur and phosphorus, biodegradation, neutralizing toxic wastes, bio-control agents and a host of other activities. Certain bacteria belonging to the families Thiorhodaceae , Chlorobacteriaceae and Athiorhodaceae contain bacteriochlorophyll and various carotenoids and are also capable of photosynthesis. Bacterial interactions with the coffee bush and the surrounding flora are known to improve plant growth and productivity. The fertility of the soil is directly dependent on the activity of soil microorganisms. The soil microorganisms mineralize insoluble and indiffusable organic constituents and make them available to plants. The common denominator to assess a healthy soil is the viable number of soil microorganisms. The shade grown Indian coffee ecosystem is unique in the true sense that it simultaneously achieves the goals of agricultural production in terms of coffee, pepper, citrus, vanilla, areca nut and cardamom production on one hand and the conservation of biodiversity on the other hand. Nothing would be wiser for the world's coffee producing Nation's to follow in India's footsteps and grow coffee under the canopy of trees. There in lies the path to sustainability. In our humble opinion we strongly feel that only shared prosperity can make the future of this planet secure. At JOE'S Sustainable farm we work with ideas that might work or might not work. But in the end analysis, these ideas are the core to the survival of Planet Earth.

Chapter 7

Prospects and Limitations of Agricultural Biotechnologies

Introductory Remarks

It is now generally that the practice of plant breeding will move forward as progress is made in knowledge and technology. Breeders in the future will not only be forced to hone their traditional skills but will also have to integrate them with knowledge and experience to coincide with more recent advancements and technologies derived from cellular and molecular concepts and approaches.

For worldwide crops such as potato, rice, maize and in the future wheat, as well as for high value vegetable crops, the future of breeders depends upon the use of a combination of breeding practices (traditional breeding, genetic engineering and tissue culture). In addition to an increasing reliance on breeding, agriculture will also depend on biocontrol to complement and to make possible the use of chemicals that are compatible with intelligent management of the natural resources essential for a sustainable yet highly productive agriculture.

Thus most aspects of agriculturally relevant biotechnology are covered. Plant tissue culture techniques not only provide essential ways for the clonal propagation of many agriculturally important crops(e.g woody plants ornamentals and vegetables) but they are also the basis for the production of transgenic plants and are at the forefront of recent studies in plant physiology, developmental biology and biochemistry. Spectacular progress in our knowledge of plant sciences, including those relevant to agriculture and breeding, can be expected from the integration of more traditional sciences with molecular and cellular concepts and techniques. A prime example is seen in the impact molecular genome analysis is already having on traditional breeding (via so called map-based breeding) but also on research and molecular breeding (e.g, via map based gene-cloning) This rosy description of the future of genetic engineering is richly tempered in this book by a realistic assessment of the many problems that still need to be faced before what is predicted can be accomplished. It was therefore wise to ask scientists well known for their "no-nonsence" approach such as Ingo Potrykus and his colleagues to write about genetic engineering of crop plants.

To quote from their introduction: " it has been possible to develop the state of the art of gene transfer to plants to a level at which many of the major crop plants are accessible to the technique and for which we have good reason to believe that there is no basic biological principle that will prevent gene transfer to a specific crop plant " Yet these authors clearly identify the considerable research and development that is still needed before one can, in many but not all instances, consider large scale applications.

A recurrent point made and well documented in various chapters of this book is that genetic engineering-because it is not limited by the natural barriers that ensure the maintenance of species and prevent transfer of genes between unrelated organisms-is an extremely powerful way to increase the genetic variability available to breeders. The present rise in interest to acquire new knowledge via fundamental research can best be illustrated by reminding ourselves of recent progress in our understanding of the molecular physiology of development, of the mechanism of action of old and newly discovered phytohotmones and of growth signals used by microorganisms that are pathogens or symbionts of plants (such as for instance lipo-chitooligosaccharides).

Biotechnology can contribute to making traditional agriculture more productive while at the same time more sustainable. It can also help achieve the goals of "non-food" and or cash crops agriculture as has been described in the various chapters dealing with crop improvement and metabolite production. Hopefully sufficient attention will be paid to regional crops of importance by local populations in developing countries.

Finally topics such as "Biofertilization" and "Bioremediation" are convincingly dealt with Topics such as "Biotechnology of Farm Animals" and "Marine Biotechnology" as well as the unavoidable but all important "Legal and Public Aspects" round out the all inclusive title of

Agricultural Biotechnologies. Some topics have already come to blossom and illustrate well the potential of present day plant sciences. There is the breakthrough in the cloning and characterization of resistance genes along with significantly improved understanding of the nature of avirulence genes and elicitors. Many hypothetical models predict that resistance genes should be involved in the signal transduction pathways linking pathogen derived signals with the regulated expression of defence genes and of programmed cell death (hypersensitivity reaction). The first available data on resistance genes show that this concept may well be correct in its very general sense, even if all resistance genes do not code for elicitor receptors.

We may safely predict that in the near future we will know more details of the molecular mechanisms underlying plant-pathogen resistance mechanisms. Several elicitor receptors will soon be isolated and characterized along with genes involved in localized pathogen-induced programmed cell death. Plant breeders will make use of isolated resistance genes to produce crops with improved resistance to a variety of pathogens and pests. The content of this book fully justifies the conclusion that Biotechnologies are bound to play an ever growing role in agriculture not only in the long term future, but also presently, and in the near future.

Fortunately the progress in applied plant biotechnology corresponds with and is in fact stimulating fundamental scientific progress. This is important because the real driving force behind this science is the success of applied plant biotechnology such as: 1) plant protection based on the expression of several different introduced genes coding for different principles acting on the same targets and thereby providing a possible solution to the problem of the emergence of resistant pathogens and pests, 2) new quality traits as well as future crops producing tailor-made non-food products (lipids, carbohydrates, and biodegradable thermoplastics). Only if plant biotechnology becomes a commercial as well as an environmental success will sufficient support be available for this fascinating research.

Plants and Agriculture

Human population has grown steadily since the advent of agriculture, nearly 12,000 years ago, which ensured a continuing and reliable supply of food. The most rapid growth occurred during the past two centuries during which time agriculture became highly industrialized, public health improved and there were increases in incomes and international trade in food grains. Despite the age-old human concerns about balancing population and food supplies and many population control measures the worlds population is projected to continue to grow into the early decades of the 21st century, stabilizing at 9-11 billion sometimes during 2030-2050. Much of this growth will occur in the already overpopulated underdeveloped and poorer regions of Africa, Asia and Latin America, which will be home to nearly 90% of the human population.

Food shortages were common during the early and middle years of the 20th century. The introduction of Green Revolution varieties of wheat, rice and maize during the mid-1960s reversed this trend and helped avoid major food shortages by keeping increases in food productivity slightly ahead of population growth. During this period, both China and India, the two most populous nations with chronic food shortages, became bet food exporters.

Realistically however such increases in food productivity can not be sustained indefinitely. It is not surprising, therefore that increases in food productivity have begun to decline during the past few years. Furthermore, improved economic conditions in China and India have created greater demand for better and varied food products, particularly poultry and meat, which require greater supplies of feed grains. As a result, both China and India have now become net importers of food. World food reserves have declined from a high of 77 days to less that 50 days.

With the current trends of population growth and agricultural production the demand for food in the most populous parts of the World will double by the year 2025, and nearly triple by 2050, Increases in food productivity of this magnitude can not be brought about in such a short period of time by conventional breeding, especially when some of out most important crops are approaching the physiological limits of productivity, not by increasing the amount of arable land, which accounts for 97% of all food production in the world..

Arable land, which is finite and comprises about 3% of the earth's surface, is deteriorating and decreasing as a result of soil erosion, salinization, over cultivation, and acidification. These factors, combined with expected increases in population, will actually decrease the global per capita arable land from the current 0.28 hectare to 0.17 hectare by the year 2025. In addition, fresh water supplies, essential for modern high-input irrigation agriculture, are becoming limited by increased human and agricultural use, and polluted by agricultural run-off and widespread use of agrochemicals.

It is feared that the resulting food shortages in the overpopulated parts of the world during the 21st century may lead to widespread social, economic, and political unrest, making food security the single-most serious threat to international peace and security. At the same time it is well known that countries with efficient agricultural systems generally have high living standards, strong economies, lower rates of population growth and democratic forms of government Increasing food productivity in a sustainable manner will, therefore, not only provide adequate nutrition to the expanding humanity, but will reduce population growth, protect the environment, promote economic development and ensure social and political stability.

The challenge for the agricultural sector during the next few decades is therefore clear double food production by 2025, and triple it by 2050, on less per capita land, with less water, under increasingly challenging environmental conditions. The situation is further complicated by the fact that in spite of the heavy use of agrochemicals, modern agriculture still loses nearly 42% of crop productivity to competition with weeds and to pests and pathogens, and an additional 10-30% to post-harvest losses due to a variety of factors, especially in the developing countries where storage conditions are poor. During the 20th century traditional plant breeding has brought about enormous increases in crop productivity. However, plant improvement by hybridization is slow, and is restricted to a very small gene pool owing to natural barriers to crossability. Beginning in the early 1980s, advances in plant cell culture can genetic transformation have overcome these barriers by making it possible to transfer defined genes into all major food crops, including cereals, legumes, cassava, potato, and many vegetable and fruits.

The entire global gene pool-whether it be plant, animal, bacterial or viral-is available for utilization. The first genes that have been integrated into crop species provide resistance to non-selective and environment friendly herbicides, and many pests and pathogens. Increasingly large acreages of transgenic maize, soybean, potato, tomato and cotton are being commercially grown for human use and consumption. Carefully planned introduction of such crops on a worldwide scale would greatly help in reducing or even elimination the enormous crop losses attributed to weeds, pests and pathogens. The use of such crops will also have a beneficial effect on the environment by significantly reducing the use of agrochemicals. Other genes for improving crop productivity, and manipulating starch/protein/oil quality and quantity, resistance to environmental stresses such as temperature and drought are also being isolated and studied. In the foreseeable future, these will be used to produce second generation transgenic crops.

It is clear that during the next few decades a wide variety of transgenic crops will become integrated into agricultural systems in the industrialized countries, Introduction of such crops into the developing countries, which need them most, will be slow and largely through the efforts of multinational biotechnology companies, because most of the developing countries currently lack the scientific and industrial infrastructure to develop and introduce these technologies into their agriculture. In the long term however, much of the increase in food production to meet the dual challenge of population growth and food demand must occur in the developing countries.

This will impact greatly on their overall economic development, which will help control the relentless increase in the population. It is critical, therefore that scientific and technical manpower and infrastructure be created in developing countries to take scientific and technical manpower and infrastructure be created in developing countries to take advantage of the remarkable advances in agricultural biotechnology. Assistance should be provided to the developing countries be international organizations such as the United Nations Educational Scienticfic and Cultural Organization (UNESCO), the Food and Agriculture Organization of the United Nations (FAO) and the World Bank through its many international agricultural research centres, like those in Mexico and the Philippines, where Green Revolution originated, International agriculture in the latter half of the 20th century was dominated by the Green Revolution, Considering the power and potential of agricultural biotechnology, it is likely that the Gene Revolution will dominate the agriculture of the 21st century.

Like the Green Revolution, it may help accelerate the rate of food productions, save lives from hunger, create livelihoods in rural households, save large tracts of land that would otherwise be needed for food production, reduce birth rates, ensure low food prices, stimulate broad-based economic growth, expand world trade, and help create a sustainable agricultural system for future generations. Finally, it should be understood that although plant breeding has long been, and will continue in the future, to be indispensable for plant improvement, it must now be complemented and supplemented by molecular breeding and genetic transformation, in order to establish a sustainable agricultural system for the 21st century.

Animal Biotechnology

Animal Biotechnology has developed rapidly from the early 1980s when the first transgenic mice and first in vitro produced bovine embryos occurred. Today animal breeding companies are using marker assisted selection to provide earlier and improved selection of breeding animals. Gene mapping efforts are annually identifying a large number of candidate genes for testing of usefulness in marker assisted selection. The propagation of genetically valuable or transgenic animals is enhanced by in vitro production of embryos.

There are already more than a dozen commercial companies selling or producing for customers in vitro produced embryos. Cloning of embryos of cattle and sheep has been shown to be possible, but low efficiencies have thus far prevented commercial use. Animal producing and genetic selection systems can be made more efficient by sexing the sperm used for in vitro production of embryos. The flow sorting of sperm cells to separate X from Y bearing cells has been successful in most species tested and in cattle and swine has resulted in offspring of the desired sex. While effective with in vitro embryos production, the sexed sperm are of less than normal viability and sperm numbers and not yet useful in artificial insemination or the freezing of sperm.

The artificial insemination and embryo transfer industries are today heavily dependent on use of cryopreserved sperm and embryos. However the cryopreservation of oocytes and the combination of some biotechnologies such as embryo biopsy and sexing with cryopreservation are still in the research stage. Transgenic cattle, sheep swine and poultry have been produced primarily by microinjections of DNA into pronuclei of eggs or by use of viral vectors. The intended use has been to change animal growth, disease resistance, wool production, or to produce new products in milk. The efficiencies thus far have been much lower than in mice. Therefore there has been little commercial use of transgenic technology except for the involvement of several companies in production of pharmaceutical products in milk.

The use of cultural embryonic stem cells and nuclear transfer to make offspring has resulted in calves and lambs. This technology offers promise of making large numbers of nearly identical offspring or more efficient gene transfer or selection. However thus far the technology is not sufficiently efficient for commercial use. Growth hormones and growth promoting factors are being developed for improved efficiency of animal growth. Growth hormones are commercially in use at present for promotion of milk production in cattle. The anatomy and physiology of birds and the late stage of egg laying prevents the introduction of DNA into pronuclei as in manuals. Thus other methods such as oocytle microinjection of DNA, viral introduction of DNA and introduction of DNA into primordial germ cells are being used in avian gene transfer research. Oocyte microinjection, blastoderm transfection and primordial germ cell manipulation have all been successful in producing transgenic birds. The production of transgenic birds appears possible but is not yet in commercial use.

Patented applications of recombinant DNA biotechnology to animal agriculture became apparent in the early to mid 1980s with development of recombinant vaccines effective against calf scours and against rabies. Today numerous vaccines in use are produced by recombinant technology. Biotechnology has also revolutionized diagnostic techniques by introducing higher sensitivity and specificity. Monoclonal antibodies, DNA probes, Southern and Western blotting, DNA and RNA amplification through PCR, constitute as integral part of the diagnostic arsenal at the disposal of the advanced laboratories.

Based on these techniques rapid animal-side field test used by the clinicians furnish results in real time, permitting immediate decisions as to the proper treatment, vaccination, or stamping-out policy. High technology based diagnosis is expected to become a routine in every diagnostic laboratory forming a sound basis for control and treatment of animal diseases.

In the short time of approximately fifteen years, animal biotechnology has moved from an unknown and unstudied science to a strong discipline of science and extensive commercial application. The science and its commercial use are rapidly progressing. The ultimate benefactor will be humans through improved food supply, nutrition and health care products available to humans around the world.

Biotechnology in Agriculture

During the last decade, tremendous progress has been made in the area of plant cell, tissue & organ culture. In vitro techniques constitute an important component of biotechnology, and have the potential not only to improve the existing cultivars, but also for the synthesis of novel plants & early release of high-yielding plants resistant to various diseases, pests, stresses & temperature.

Green revolution of recent years has been the result of efforts that led to the improvement & better selection of species of agronomic interest. Some specialists feel that biotechnology is the second phase of green revolution, studies are already in progress to obtain crop plants having higher photosynthetic efficiency, improved fixation of atmospheric nitrogen, increased nutritional quality & greater tolerance of salinity, alkalinity, acidity, deficiency of nutrients & stresses.

Wide hybridization through embryo & ovule culture both at the interspecific & intergeneric levels have been achieved in a number of crops through the culture of the hybrid ovules & embryos. In vitro fertilization techniques have also been employed to overcome incompatibilty. Haploids produced through anther culture & their diplodization have resulted in the release of varieties in crops such as wheat & rice.

Biotechnology

The term biotechnology is composed of two words-bio (Greek-bios-means life) and technology (Greek technologia-means systematic treatment). In other words, it is the science of applied biological processes.

Accordingly to the oxford Dictionary, technology is the "Scientific study of the practical or industrial arts". Spinks (1980) defined biotechnology as, "the application of biological organisms, systems or processes to manufacturing & service industries". European Federation of Biotechnology (1981) defined biotechnology as, "the integrated use of biochemistry, microbiology & engineering sciences in order to achieve technological (industrial)application of the capabilities of micro-organisms, cultured tissue cells & parts thereof".

In the report Biotechnology: A Development plan for Canada (1981) biotechnology is defined as "the utilization of a biological process, be it microbial, plant or animal cells, and their constituents, to provide goods & services".

Production of Pathogen-free Plants.

Plants traditionally being vegetatively propogated are systemically infected with viruses & other pathogens which greatly reduce yield & also Q uality of the marketed commodity. Virus diseases like potato leaf roll virus (PLRV) or potato virus Y (PVY) for example, can cause upto 95 % reductions in the tuber yield of potato crops.

Increase in yield upto 300% (Averaging 30%) has been reported following replacement of virus-infected stock with specific pathogen free plants. Since majority of viruses infect plants in a systemic manner their elimination may be achieved through meristem tip culture. Plant species for which virus-free plants have been obtained.

Species Virus Eliminated

1. Allium sativum (Liliaceae) Garlic mosaic virus
2. Brassica Oleracea (Brassicaceae) Cabbage black ringspot virus, Turnip mosiac virus, cauliflower mosiac virus
3. Chrysanthamum sp. (Asteraceae) Chlorotic mottle, Green Flower, stunt vein mottle, virus B.
4. Dahlia Sp. (Asteraceae) Dahlia mosaic virus, Tomato Aspermy, vein mottle, virus B.
5. Glycine max (Fabaceae) Soyabean mosaic virus. Impomoea batatas Feathery mottle, Hanmon Mosiac, (Convolvulaceae) ' Rugosa mosaic
7. Lilium Sp. (Liliaceae) Cucumber mosia virus
8. Monihot Sp. (Euphorbiaceae) African cassava mosiac
9. Musa sp. (Musaceae) Cucumber mosiac virus
10. Nicotiana tobacum (solanaceae) tobacco mosaic virus
11. solanum tuberosum(solanaceae) potato virus.

Although the apical meristems are often free of viruses, this cannot be regarded as a phenomenon of u niversal occurrence. The success of meristem tip culture depends upon several factors. One of the most important factors is the relative distribution of viruses in the growing tip of donor plants. There are some viruses which invade the growing tip viz, TMV, potato virus X, cucumber mosaic virus. In such cases virus free plants are obtained by combining meristem-tip culture with high temperature treatments.

Production of Disease Resistant Plants

This is one of the most useful applications of tissue culture in crop improvement. In potato, somaclones have been screened for both late & early blight resistance. In maize, somaclonal variation has induced resistance to race T of southern corn leaf blighrt. In sugarcane, resistance to diseases like fiji & downy mildew have been recovered. In lucerne, selection of cell lines & plants resistant to the toxin of *Fusarium oxysporium* has been accomplished.

Improvement of Nutritional Quality

One of the major sources of protein for human & animal commnsumption is constituted by the proteins contained in seeds of many plant species. The cereals & legumes which are major sources of storage seed proteins, contain limited amount of certain amino acids which are essential for human beings. Majority of these cereals are deficient in lysine whereas legumes are deficient in sulpher amino acids.

A wide range of approaches have been employed for improving nutritional quality of various crop plants. Important among them are selecting cell lines resistant to amino acid or their anologues of lysine, tryptophan, proline & phenylalanine. Isolation of variants over producing specific amino acids in culture has been successful, but expression in the whole plant & especially in the seed has not yet reached the level required to make an impact on protein quality.

Selection for Salt & Draught Tolerance

Continued efforts to increase intensity of cropping for increasing production from limited land resource by extending irrigation facilities have resulted in the gradual build-up of salt concentrations in the soil. This has resulted in loss of productivity on such soils. Salt tolerant lines have been produced in crop plants such as tobacco, tomato, cereals. Salt tolerance has been incorporated into rice lines with improved plant type. Some of the improved rice cultivars viz., IR 42, IR 43 & IR 52 are salt tolerant.

Development of cultivars tolerant draught can contribute significantly in agriculture economy. Tolerance to draught is a polygenic trait & involves highly complex osmo-regulatory functions. In tissue cultures, simulated draught conditions have been achieved through incorporation of non-penetrating osmotic solutes such as PEG (polyethylene glycol) and dextrans in the media.

Production of Genetically Variable Plants

The success of any crop improvement programme depends on the usable genetic variability in the base population cells in culture offer an excellent systems for inducing variations & regenerating pure mutant types. Genetic variation can be an option to lessen our reliance on cost intensive germ-plasm collection & conservation programmes.

Variant producing capacity of cell culture can be augmented to a great extent by employing physical & chemical mutagens. Somaclonal variation has been extensively exploited for the improvement of a sexually propogated crops viz., potato, sugarcane. From cell cultures, some superior cultivars have already been produced in sugarcane which are high yielding, drought resistant & temperature tolerant.

Biofertilization

Molecular nitrogen in the atmosphere is converted into Biologically converted forms by nitrogen fixing micro-organisms e.g. Rhizobium.

The most sophasticated approach to biofertilization is to create plants that possess the genetic capacity for nitrogen fixation. Attemps are being made to transfer genes for nitrogen fixation (hifgenes) from bacteria to plants.

Rapid Clonal Propagation

Tissue culture has found its best commercial application in production of cloned plants at a very high rate as compared to conventional methods. It is important specially for initially building up of propagation stock of elite clones or individual plants which are otherwise slow to multiply. A number of agriculturally important plants have been clonally multiplied.

Genmplasm Storage

The primitive cultivars & wild relatives of crop plants constitute a pool of genetic diversity which is invaluable for further breeding programmes. There are over 20,000 plant spices which are rare or threatened with degradation of their neutral habitats. The most economical form of storing germ plasm for seed propagated species is seeds. However, there are certain limitations of this method. Therefore, various methods of invitor storage of germplasm are thus of great practical significance for long term storage of germplasm. Presently, there are two approaches to invitor germ plasm storage. Slow growth technique & cryopreservation.

Biological Control of Agricultural Pests

Insects consume nearly one-third of human food supplies on earth. One of the major goals of biotechnology is to develop target specific biological pesticides that will kill some specific pest but will not harm other species. Several insect-specific pathogens, developed through biotechnological processes, are being produced commercially for their use as microbial pestices.

Bacillus thuringienis inocula have been used as bio-insecticides. Toxin produced by the bactenium kills the gypsy moth. The prevention of crown gall disease formation is achieved by spraying young plants or seeds avirulent species *Agrobacterium radiobactor* var, *radiobactor*. Microbial insecticides have a number of advantages over traditional chemical insecticides:

a) they are specific for a small number of species of insects & do not kill plants, animals & beneficial insects.

b) they are cheaper than organic pesticides and

c) they do not leave any residue effect.

Selection for Herbicide Resistance

Herbicides are used to control weeds in agricultural fields. Tissue culture techniques offer several advantages in developing herbicide-resistance plants:

1. The techniques for selection are relatively rapid;
2. Small volumes of cells represent the genetic potential of hectares of fields.
3. Metabolic studies are facilitated by uniform & sterile conditions &3
4. Possible new types of resistance not found in nature may be obtained from plant cell cultures.

Attempts to produce herbicide tolerance using somaclonal variation & invitor selection have already been made.

Biotechnology and Indian Agriculture: The Challenge of the Next Millennium

The world has won the important battle in the area of food security, but the war is still on. A total of 800 million people, that is one out of every six persons, in the developing world do not have access to food. One-third of all pre-school children in the developing countries are food insecure. We, thus, have a big challenge ahead of us as we enter the next millennium. It is true that the mass starvation that was predicted for Asia in seventies and eighties did not occur. It is only because Science was effectively put to work to raise agricultural productivity. In India, the '*Green Revolution*' was a success due to the introduction of improved seeds, fertilizer, irrigation and plant-protection measures combined with positive policy support, liberal public funding for agricultural research and development, and dedicated work of farmers. Notwithstanding all-round achievements, the basic problems of food security, poverty, equity and sustainability, continue to be a cause of concern in India today.

The Indian Challenge

There are several concerns about Indian agriculture that we need to address. Even in the green revolution crops, rice and wheat, the Indian average yield ranks around 50th in the world. A weak agro-based industry, the problems of weak marketing, storage, transportation, credit support, etc, haunt us. Only 0.3% of agricultural GDP is spent on agricultural R &D. The population is growing at an alarming rate of 1.8% per annum. It will be around 1.3 billion by 2020 and would require 50% additional foodgrains. A total of 200 million people are still below the poverty line and have insufficient access to food. Declining resource base, degeneration of natural resources, including soil fertility, water environment, etc, are some other concerns.

Food production systems centre around a few states only. Out of 17 big states, only 7 are self-sufficient and surplus in foodgrains. There was a significant decline in per capita intake of calories and protein between 1975 and 1990. Inspite of all these problems, there is plenty of good news. India harbours a vast diversity of basic natural resources. Tremendous bio-diversity of agriculturally important micro-organisms exists, which are not yet been harnessed. We have a variety of crops, including quality fruits, vegetables, foodgrains, and a number of plantation and commercial crops. Indian fisheries resource has spread over 8,100 km of coastline on the east and west, and provides access to marine resources and transportation.

Very large proportion of area of various food crops is under low productivity category. It can be increased with inputs of high class science. There is a scope to cultivate sizeable portion of waste lands through soil amendment and introduction of agroforestry. Several million hectares of saturated soil can be potentially exploited from rainfed lowlands of eastern India.

The India Exclusive Economic Zone (EEZ) of the marine sector, of about 2 million sq.km., has high potential harvestable yields. Only about 20% of the cultivated area is covered under tractor cultivation, great scope of making quantism quantum jump in production and productivity exists, changing consumption patterns offer scope for diversified agriculture. Trade liberalization provides opportunities to reach for outside markets, which were not accessible otherwise. Great gains are possible through agro-based industrialization and overall rural development.

The cultivators and farmers of India have a rich heritage of traditional agricultural wisdom on crop and livestock husbandry and fisheries; based on the principles of the conservation of natural resources and environment. India has an extensive network of extension channels for dissemination of technologies generated by research institutes. India has a strong manufacturing capability and a large network of about 20,000 manufacturers and nearly one million village artisans for indigeneous production of all types of agricultural machinery. Therefore, inspite of several disadvantages, we have a great chance to leap forward, if we enforce the right policy initiatives and implement them.

Biotechnology and India Agriculture

We need to recongnise that, just as the food reequirements of today's population of nearly 6 billion people could not have been met by the technologies of the 1940's, we cannot assume that current practices will feed the population of 8 billion expected by 2020. New approaches are needed in addition to the continued improvement of existing methods of crop and animal husbandry and food processing.

The strategic integration of biotechnology tools into India agricultural systems can revolutionise Indian farming and usher in a new era. Genetic engineering is clearly the most revolutionary tool to impact agricultural research since the discovery of genetics by Mendel. Prior to genetic engineering, the exchange of DNA material was possible only between individual organisms of the same species.

With the advent of genetic engineering in 1972, scientists have been able to identify specific genes associated with desirable traits in one organism. For example, a gene from bacteria, virus or animals may be transferred into plants to produce genetically modified plants having changed characteristics. This method therefore allows mixing of the genetic material from species that cannot otherwise breed naturally.

Genetic engineering can be vital for an agrarian country like India. It can help in minimising the crop damage through disease and pest resistant varieties, reducing the use of chemicals, enhancing stress tolerance in crop plants, thus permitting productive farming on unproductive lands, etc.

One could even extend the growing season of crops and minimise losses due to environmental factors on the one hand and increase the shelf life of fruits and vegetables, on the other, thus minimising losses due to food spoilage. We can expand the market vista and improve food quality. Biotechnology can also produce plants that possess healthy fats and oils, possess increased nutritive value, and create a whole range of higher value feeds. Biotechnology has even the power of producing biodegradable plastics, edible vaccines, etc.

It is only through the blending of the 'gene revolution' with our experience in the 'geen revolution' that we can reach our goal of 'evergreen revolution' and also 'nutritional revolution' The advantage of the gene revolution is that it is relatively scale neutral, benefiting big and small farmers alike, It is also environment friendly. Thus, it can be of great help to the smallest farmer with limited resources in increasing farm productivity through the availability of improved but powerful seed. It can also reduce a farmer's dependency on chemical inputs such as pesticides and fertilisers.

Modern biotechnology offers unlimited opportunity for enhancing genetic potential of crops and other commodities, management of biotic and abiotic stresses, bio-remediation and organic recycling. India is one of the main centres of agricultural biodiversity and its gene richness can greatly complement the developments in modern biotechnology. Likewise, new developments in GIS, remote sensing, and crop modelling, provide new opportunities for integrated management of natural resources.

The revolution in informatics provides opportunities for sharing latest information for research and planning in a highly organised and efficient manner. With the globalisation of economy, the opportunities for value addition and post harvest management are also immense. With clever blending of technologies, the India farmer can usher in new era of confidence and performance. Sir Francis Bacon had once said: *'It would be an unsound fancy to expect that things which have never yet been done, can be done except by methods which have never been tried'*

In the new context, we have to review the new role of biotechnology. Also, it is in this context that we have to view the rapid advances that modern biotechnology is making, including the area of transgenics. Let us focus on some of the developments in this area, since considerable interest and controversy has been created around the world today in this exciting area.

Movement in Transgenics

The advent of genetic modification over the last two decades has enabled plant breeders to develop new varieties of crops at a faster rate than was possible using traditional methods, with huge potential for further beneficial development. Tobacco was the first plant to be genetically transformed\, in 1983, with cereals beginning in 1990, but it is only recently that products such as GM soya have reached the market place in Europe, and, for a variety of reasons, given rise to concern & controversy.

The concern must be addressed seriously if society is to exploit the new technologies appropriately. We will address this issue rather extensively a little later. Whereas in India, we have been discussing and debating the issues, the rest of the world is galloping ahead.

Table: *Ag biotech product sales*

$ Millions	*Sales*			*Annual Growth Rate*	
	1994	1997	2002	1994-97	1997-02
Transgenic seeds and plants	20	405	2,030	173%	38%
Animal growth horomes	80	225	405	31	12
Biopesticides	35	65	110	23	11
Other	107	180	340	19	14
Total U.S. Sales	$242	$875	$2,885	53%	27%

Development, field testing, and commercialisation of transgenic crops, is now recongnized by the international scientific community to be an essential strategy for food security. Fortyfive countries, including India, have conducted transgenic crop field trials up to 1997. The activity in the developing countries has been shown in table below.

Table: *Releases of transgenic plants in developing countries (by species and introduced trait)*

Species	***Introduced Trait***					
	Total Fied releases	***Herbicide Resistance***	***Insect Resistance***	***Virus Resistance***	***Product quality***	***Others***
Maize	46	29	16	1	3	5
Soya bean	27	25	—	1	1	—
Cotton	24	16	15	—	—	—
Tomato	19	—	2	1	16	—
Potato	13	—	1	6	—	6
Subtotal	129	70	34	9	20	11
Other species	30					
Total	159					

The emphasis has been mainly on traits for herbicide, insect, virus & fungal resistance, and on product quality with a limited number of genes. Out of a total 159 releases. Argentina leads the way with 43, followed by chile, Mexico, Puerto Rico and the Republic of South Africa, with 17-20 each. India is far behind with just a few releases. There are fears relating to transgenic plants in society.

These need to be addressed, not by brushing them aside, but by careful scientific analysis and educational programmes. All biosafety protocols must be followed rigidly. Let us deal with this issue more exhaustively now.

Public Concerns

The potential of biotechnology as a method to enhance agriculture productivity in the future has been accepted globally. However, because of its revolutionary nature, risk and uncertainty may be created by the process of genetic engineering and by the resulting genetically modified products. This may result from the introduction of a new unrelated DNA sequence into a recipient organism.

The introduced DNA may have some unexpected effects on the cellular processes of the recipient organism. In addition to this, introduction of antibiotic resistance genes, as selection markers, pose serious implications in public health especially in genetically modified plants that are directly used for food purposes.

Risks are also associated with the genetically modified plants that are released into the environment. The nature of interactions with other organisms of the natural ecosystems cannot be anticipated without proper scientific testing. For example, modified plants with enhanced resistance to pests or disease threaten to transfer resistance to the wild relatives, which will have implications for biodiversity and ecosystem integrity. These, and many more, doubts plague the minds of common people.

Therefore, it is very clear that if the biotechnology potential has to be completely realized, it has to be done in an extremely responsible manner. At each step, proper testing and scientific data have to provided. Not only this, proper regulatory and policy mechanisms have to be put into place. It is therefore, absolutely essential, that conscious efforts be made especially by the developing countries to initiate well-defined programmes for the development and regulations of genetically modified plants. To make this possible and to benefically, ethically and sustainably reap the benefits of biotechnology, there is a pressing need for scientists, researchers, policy makers, NGOS, progressive farmers, industrialists, and representatives of the government, to come together on a common forum in order to discuss the common concerns and find solutions to them.

Will modified plants transfer their introduced genes into wild relatives growing nearby?

Will modified plants that produce new compounds disrupt the 'balance of nature' in some way?

Could the planting of a restricted number of cultivators leave crop plants more susceptible to disease?

Could the planting of a restricted number of cultivators lead to a reduction in biodiversity (of crop plants, weed species, insects and microflora in the fields in question)?

Will genetically modified plants be able to avoid the factors that regulate natural populations and thereby change the usual ' balance' between populations?

Similarly, there are concerns about the social and economic effects. For example, some of these concerns are:

- How will the structure of farming (particularly in developing countries) be affected by biotechnology?
- How will patent laws affect traditional breeders 'rights (e.g. the right to save seed from one year to the next)?
- Will plant breeding be left increasingly in hands of a few companies, and if so, what effects might this have?
- Will some countries be 'plundered' for their genetic resources?
- We also have a set of ethical and moral issues that need addressing. In particular,
- The consumer has the right. What will consumers be told about the new food products?
- Is it acceptable to 'interfere' with nature through genetic engineering?
- Do we have the right not to use all means available to improve crop plants, especially when so many people are under or malnourished?

Then there are several regulatory issues. Among these, the three key issues are:

- Do current regulations give sufficient protection to farmers, consumers, those who have invested in, and those engaged in research?
- Is there sufficient international legislation to ensure environmental protection?
- Do current regulations compromise the competitiveness of biotechnology companies by being excessively restrictive?

The Royal society of London had appointed a group of experts to examine various aspects including the scientific evidence concerning the risk of transfer of genes from Genetically Modified (GM) crop plants to wild species and non-GM crops, the uptake of genes from GM food by the digestive system, and the current state of the regulatory system.

The experts concluded that the chances of gene transfer happening are slight, provided the regulatory processes are followed, but that this must be kept under consideration.

The experts also concluded that the uptake of genes via the food chain is not a new issue because genes (i.e. DNA) are normal constituents of the human diet. Many products from GM Plants, such as sugar prepared from GM tomato paste, are so similar that they are regarded as 'substantially equivalent' others, for example flour from GM soya, may contain a new gene or its product, although many of the purification processes involved in food production will destroy and DNA present in the raw material.

Some GM foods have been produced using an antibiotic resistance 'market' gene, which is a laboratory device designed to identify genetically transformed plants. The Royal Society has shown concern about the use of such genes in food products and it suggested that any further increase in the use of such markers in the human or animal food chain would be undersirable. Gm crop plants have been produced to improve insect tolerance and virus resistance, and to include herbicide tolerance, so that transferred to non-target species of plants, and that the development of resistance by target pests is minimised, the regulatory authorities must be assured that any negative effects would be no greater than those resulting from conventional procedures; and that any long term effects on the environment and ecology would be closely monitored, with statutory restrictions in place to control marketing; and also that best practice advice was adopted by growers.

The reliance on a case by case approach to the legislation, may result in lack of analysis of the overall impact of the technology on agriculture and the environment, and of the long term effects of gm. Although mechanisms are already in place to regulate many individual aspects of GM technology, there is no means for looking at GM technology as a whole. As independent, overarching, regulatory body is needed to span departmental responsibilities, monitor the enforcement of existing or future regulations, and strengthen the guidelines to growers of GM crops, such as those specifying the isolation distances between Gm and non-GM crops. The body should also review and monitor the membership of advisory committees and regulatory bodies.

The Indian Effort in R & D

Indian has set up a large network of research institutions and agricultural universities located in different regions and states of the county.

Biotechnological Applications Relevance to The Indian Farmer

Increased food production must keep pace with the country's population increase. Not only the quantity of food made available should improve.

To keep pace with the present population growth and consumption pattern, India's food requirement has been estimated to cross 225 million tons by 2005 AD.

This would mean an annual agricultural growth rate of 6.7 percent, a daunting task considering the rapidly shrinking resource-base and fast declining input-use efficiency in major cropping systems. Not withstanding the impressive gain in the agricultural production, the vast agricultural potential still remains highly under-utilized.

There are serious gaps both in farm yield realization and technology transfer, as the national average yield of most crop varieties of most crop varieties is low.

Not many superior varieties is mot pulses, many oilseeds and vegetable crops have been evolved in recent past. In case of vegetables the varieties which were imported or developed many decades ago are still being grown by the farmers.

Their yields are low and their inherent ability to respond to improve production technology is also limited. Farmers have accepted bybrids in my vegetable crops such as cabbage, tomato and capsicum.

Research and Technology Transfer

The Central and State governments continued to invest heavilyu in research, hence the Indian public sector research system has increased in size in real terms but by mid 90s levelled off. The number of hybrids and varieties developed and released by ICAR Crop Improvement Projects and State Agricultural Universities in major and minor crops are enormous and have been well documented.

Private research to develop superior hybrids was initiated by few seed companies in mid 60s.

By 1987-88, the number increased to over 12 and in 1997-98 it is estimated at 34.

With the governments recognition of provate sector's performance capabilities and with "New Seed Policy" announced in 1988, the number of private companies involved in crop improvement programs have been steadily increasing.

Number of Private Companies Engaged in Crop improvement Program (1987-1998)

Table : *Number of Companies with Crop Improvement Programs*

Crop	*1987-88*	*1993-94*	*1997-98*
Pearl Millet	12	23	25
Sorghum	10	12	15
maize	06	15	20
Sunflower	10	30	20
Cotton	09	21	34
Hibrid Rice	-	4	6

For want of financial, infrastructural and trained human resources, most private companies specialize in breeding superior hybrids in few crops characterized by low volume, high value and amenable to hybrid technology. In addition to the ICAR and Agriculture University research system, about 13 private sector seed companies are actively involved in vegetable R&D. Some of them have joint ventures with multinational or companies specializing in vegetable crop research and utilize research facilities available with their overseas partners. Few have created facilities for field testing imported and public bred varieties. The role of private sector in vegetable research is primarily in the development of F hybrids and few open pollinated varieties.

Table: *Hybrid Vegetable and Melon seeds of Private R&D available in the market (1996-97)*

Crop	*No. of Companies*	*No. of hybrid Varieties*
Tomatoes	11	45
Brinjal	10	38
Chilli	8	27
Bhendi	13	25
Cabbage	11	40
Cauliflower	8	21
Ridgegourd	6	8
Watermelon	5	13
Muskmelon	2	7

Number of private/public hybrids in major crops in market during 1995 are compared with the number in 1997.

Table: *Number of private and public bred hybrids in major crops in the market during 1995 and 1997.*

Number of Hybrids in Market

	1995		*1997*		
Crop	***Private***	***Public***	***Private***	***Public***	***No of Private Hybrids grown on 2% plus area***
Pearl Millet	50	04	60	06	14
Sorghum	22	04	41	05	6
Maize	57	03	67	03	12
Sunflower	47	05	35	06	10
Cotton	73	15	150	15	19
Hybrid Rice	4	4	8	4	-

In pearl millet, sorghum, maize and cotton number of hybrids introduced by private, companies has been steadily increasing. Many hybrids especially in paddy and cotton have regional adaptability. The number of hybrids which are accepted by the farmers and may be grown on 2 percent or more area is much lesser, so also the number evaluated in coordinated project trails or identified for release. Many hybrids recently supplied by the private sector embodies new technology which is either imported or developed by private companies. Private R&D's real investment in research has quadrupled between 1986 and 1997. Subsidiaries or joint ventures with multinational companies account for 30 percent of all private seed industry research in 1997. Both domestic companies and multinationals companies have started research programs on rice and mustard. Recently, Monsanto has also focused on hybrid wheat.

Few private sector companies have also initiated joint ventures (e.g. Mahyco-Monsanto Biotech, ProAgro-Agro-Evo) to transfer specific genes to their product lines to provide effective protection against pests. Some of them have taken initiative to treat their seed with high-tech chemicals to provide protection against sucking pests (aphids, jassids, thrips) during seedling and grand growth crop stages and diseases (downy mildew in bajra) during flowering.

Commercial Seed Production

Majority of certified/commercial seed is produced under contract with trained seed multipliers in suitable agro-climatic conditions. Both public and private companies make contact with progressive farmers or trained seed multipliers and negotiate terms and conditions of production. Individual seed multipliers have small areas to contract and this greatly increases the cost of production and administration. To overcome this situation, contracts are arranged through local organizers, who in turn contract individual farmers, consolidate individual fields intoblocks and supply to the company the ultimate product.

Companies generallyprepare a technical package for seed multiplication specifying the kind of soil needed, isolation distances required, staggered plantion in case of a hybrid if needed, distances among rows and plants, depth of planting, time of nutrient and spray applications, frequency and time of irrigation, mode of harvest especially in case of hybrid seed etc. They prefer to contract between 300-400 acres in a village and designate such an area as a "seed village". They create seed processing facilities and place technically trained staff to guide seed multipliers, monitor actual production and supervise seed processing. Generally one such unit serves a cluster of 5-10 seed villages

Only seed lots meeting genetic Purity, Physical Purity, germination and disease standards prescribed by the government of India, are packed and commercially sold. User-farmers have become very conscious not only of genetic purity and germination, but also of physical attributes of seeds. They prefer bold, shining and plump seed, unaffected by mould The companies therefore, organize seed production in areas with appropriate agro-climatic conditions and no rains in the post-flowering period of seed crop. They also ensure proper cleaning to remove undersized, premature, broken seeds or inert matter on cleaner-cumgraders, gravity table, de-stoners and thus improve physical attributes of seed lots. Seeds are treated with recommended pesticides to prevent them against stored grain and soil pests. To add value to their seed, many companies now treat them with specific chemicals to provide protection agains sucking pests during seedling stages or specific diseases such as downys mildew for 40-45 days after emergence.

The seed production of most hybrid crops is generally undertaken in Western and Southern parts of the country because of congenial agro-climatic conditions and possibility of taking two seed crops in a year if necessary.

The prominent states are Gujarat, Maharashtra, Andhra Pradesh, Karnataka and Tamil Nadu. In contrast, the Northern states from Punjab to West Bengal are dominated by rice, wheat, and legume crops which are less profitable. Consequently there is a major difference in the cadre of seed industry between the North and West/South of India.

Technology of seed production is moving at a rapid pace. Hence the companies have to ensure that they pass on the technology to the contract seed multipliers to harvest high yields of good quality seeds. Most public bred hybrids and variety seeds are certified while the proprietary hybrid seeds are sold under truthful label. All seeds meet minimum prescribed standards but for proprietary hybrids quality monitoring is through their in-house system.

Private Hybrid's Share of Seed Market

Private sector has played a major role in supplying quality seeds of both private and/or public hybrids. It is interesting to note that 69% of the pearl millet seeds produced during 1996-97 season was of private hybrids. Other crops where private research plays key role in seed supply are sudangrass sorghym (100%), sunflower (98%), Maize (90%), sorghum (52%) and cotton (50%).

The contribution of private research in value terms is also steadily increasing. The share of research hybrids to total turnover in these crops was about 70 percent in 1996-97 as compared with 46 percent in 1990-91. Private hybrids are successfully meeting the regional needs of early maturity, tolerance to pests, unpredictable agro-climatic condition and soil variations.

Government Intervention: Legislation and Policy Issue

At the highest level, government policy is to reduce controls and to make the economic system more transparent in practice however, there is potential proliferation of legislation affecting seed sector. Given the current climate of economic development there is an urgent need for positive relationship between the government and the private sector. In order to fulfil farmer's needs and GOI food production targets set for the year 2005, the existing misapprehensions should disappear to ensure private sector investment in infrastructural improvement and expansion to increase quality seed production.

Presently Indian seed industry is governed by the following Acts:

- Seed Act 1966
- Seed Rules 1968
- Seed Control Order 1983
- Essential Commodities Act 1955
- Package Commodities order 1975
- Standrads of Weights and Measures Act 1976
- Consumer Protection Act 1986.

Export Regulations and Quarantine :

- Plants, Fruits and Seeds (Regulation of import into India) Order 1989
- Urban Land Ceiling Act 1976
- State Acts of land acquisition/land use
- State Acts for the control/movement of crops and seeds.

The policy framework hitherto has been regulatory with excessive legislation. Above plethora of legislation covering the seed sector is out of date and restrictive rather than progressive. The multiplicity of Acts, rules and administrative orders have resulted in over regulation of the nascent industry.

Future Trends

During past four decades spectacular progress has been achieved in agricultural production in the country. Compared to 51 million tons of food grain production in 1950-51, India realized a food grain production of over 196 million tons during 1997-98, resulting not only in self-sufficiency in food but also in a buffer stock of over 30 million tons. Food grain production is expected to cross 200 million tons in current year (1998-99). In the oilseed sector, which is primarily production was coupled with important policy planning and research and technology development transformation. The steady increase in productivity has been accomplished by exploiting hybrid vigour in maize, sorghum, pearl millet, sunflower, cotton, castor and improved genetic potential of high yielding varieties in wheat, paddy, many legumes and oilseed crops. The superiority was for high inherent harvest index, inbuilt resistance to major disease-pest complex, ability to withstand environmental stress conditions, adjusting growth cycle to better suit the available growing conditions, generally greater physiological efficiency and/or improved quality. Nineties focused on exploiting heterosis in rice, rapeseed-mustard and cajanas for enhanced productivity, sustainability and utilizing the vast agricultural potential which still remains highly under-realized through bridging the serious gaps in inherent yield potential and ensuring a rapid technology transfer to increase national average yields leading to increased productivity. The Indian seed industry shows a clear pattern of development that also reflects in government policies. The decade, after the announcement of New Seed Policy of 1988, witnessed entry of major multinational companies and foreign genetic material to the Indian seed market. Nineties also saw proliferation of domestic seed companies into more dynamic prevailing economic climate. Although public sector dominance continues in open pollinated variety seeds, the private sector is steadily increasing its market share of hybrid seeds. Commercially the seed industry is going through a period of accelerated development with activities now including in-house research, adoption of agro-biotechnology and genetic engineering techniques.

Till such time that the population stabilizes, the demand for food would continue to increase. To keep pace with the present rate of population growth and consumption pattern, India's food requirement has been estimated to cross 225 million tons mark by 2005. The certified/ quality seed distribution has considerably increased from a mere 18,300 tons in 1952-53 to over one million tons in 1997-98. The quality seed requirements are estimated to increase to 1.8 million tons by 2005 AD. If the population growth continues at current rate, the country could require 300 million tons of food to feed an estimated 1.4 billion people by the year 2025. This will only be possible with an average agricultural growth rate of about 5 percent annually. India, therefore, need to continue the green revolution through introduction of value added superior genetic material possessing inherent high yield potential even under environmental stress conditions and genetically engineered or chemically induced seed treated tolerance to major disease-pest complex. Then only the income levels of small and marginal farmers who dominate Indian agriculture today will rise. Accomplishing such targets of production will not be an easy task considering the fast declining input use efficiency for major cropping systems and rapidly shrinking resources base.

Future agricultural development would be guided not only by the compulsions of improving food and nutritional security but also major concerns for environmental protection, sustainablity and profitability. Following GATT agreement and the liberalization process, globalization of market would call for competitiveness and efficiency of agricultural production. Projected population migration from rural to urban areas will also demand a significant shift in the pattern of the food basket. Developments in biotechnology, molecular biology including genetic engineering and new cutting edge technologies in recent years will provide new potential for accelerated growth. Used in consort with conventional approach and harnessed judiciously, these new break throughs would help increase agricultural productivity and sustainablity.

To meet the future challenges of increased productivity-a daunting national need, the problem oriented, demand driven, value added approach of integrated research and development involving all reasonable players as partners would offer practical solution. This would necessitate an active cooperation between the public and private solution. This would necessitate an active cooperation between the public and private sector agricultural research, massive private investments in R&D, global exchange of germplasm and technology and necessary changes in government policies to create a congenial environment to encourage entry and active participation of technology rich foreign private companies in Indian seed market. Investments and technology would needed to breed hybrids and varieties with high inherent genetic yield potential with value added traits, incorporated through genetic engineering to improve nutrition, provide protection against major pest-disease complex, inbuilt drought tolerance to ensure increased productivity even for arid and semiarid zones of the country. Biotechnology applications would not only minimize loss of biodiversity but will also expedite introduction of high yielding traits in desired crops of major economic importance.

ICAR is currently preparing a new perspective plan to upgrade farm research and man power development systems in the country. They are being reoriented to ensure sustainability and globalization of agriculture with emphasis on diversification, value addition and to encourage exports. This is opportune time for bold initiatives and policy decisions in support of private R&D which is steadily becoming stronger. The number of proprietary hybrids is increasing rapidly. Hybrids in rice are being commercialized. Hybrids in rapeseed-mustard and cajanas are on threshold of commercialization. Research efforts to evolve hybrids in wheat are on. Innovations in genetic engineering are being incorporated in Indian products; in cotton to provide resistance to Heliothis bollworm; in brinjal, cauliflower and cabbage among vegetables for black moth. Genetically engineered products will probably enter commercial market in early next decade. The present value of formal seed market estimated at Rs. 20,190 crore, i.e. US$ 480 million with private sector contributing approximately 72 percent, is likely to double in next 3-4 years. There is lot to be gained from these synergies that remain untapped between government laboratories, industrys and academia. It is an appropriate time for an effective partnership between ICAR projects and institutes, state agricultural universities and private R&D to evolve.

Time is ripe for a second green revolution which would not only concentrate on increased productivity but also on value added traits to reduce cultivation cost, pollution dangers and improve quality of products. Rapid developments in biotechnology, particularly in genetic competition. Indian seed industry is ready to face the challenge. Government is also cognizant of need to remove bureaucratic hurdles to its expansion. It is most opportune time for government to take bold decisions in favour of private sector development. It needs to start recognizing and rewarding innovations and encourage domestic companies to invest greater resources in research. Amended Indian Patents Act Of 1910 includes biotechnology and genetically engineered product patents and has been Passed by Rajya Sabha (Upper House). Government should expedite implementation of Plant Variety and Farmer's Rights Protection Bill. These steps would send right signals to the international of Plant Variety and Farmer's Rights Protection Bill.

These steps would send right signals to the international seed market and would encourage massive investments both from multinational and domestic seed companies to set up high-tech R&D, seed production and conditioning facilities to produce seed meeting international standards, for both domestic market and for export. Seed industry needs to become technology driven and innovation led. This transition will have to be facilitated through innovating strategies where both government and industry are partners. The private sector's R&D emphasis on hybrids would also then include open pollinated varieties in major self-pollinated crops where hybrid systems are not yet identified. As the business expands, large companies would create a full line of products to meet their dealer's demands. Since most of this technology will be high value proprietary products, many farmers are likely to become contract growers. Their risk will be limited, and profits although smaller will be guaranteed. New technology thus holds the potential to make farming more secure and profitable.

The guiding philosophy of ICAR research for next millennium's hi-tech environment should be to complement private sector R&D efforts. ICAR's "research product mix" should reflect priorities and targets that continuously evolve in response to changing internal and external research environment. Its policies should consider a shift in emphasis to upstream research producing mainly intermediate outputs, i.e. inputs to further research, which in turn will yield better products. In past 4-5 years over 19 billion US dollars have been committed to mergers, acquisitions and technology alliances in the crop genetic industry world wide. There has also been incredible growth of big company R&D budgets in this area. The agriculture in the developing world is becoming like any other industry and is being dominated by a hand full of big technology developers. Until a decade ago hundreds of competing companies except for one or two big ones accounted for nearly half of total seed market share. Now the technology is a hand full of large companies developing the technology and controlling much of the industry and fewer smaller players. This worldwide changes would have a major impact on the pattern of Indian seed industry growth in the coming century.

Seed industry worldwide is also going through a rapid technology revolution with value added output traits at its centre, driven by consumer and commercial end users. It involves all aspects of agri-biotechnology and crop production. Emphasis is on proprietary genes in value added crops for the feed, food, chemical and pharmaceutical industries at the centre of new seed industry.

This change to value added agriculture will also have major impact on restructuring of Indian seed sector, hence seed business in coming decades. It will become more difficult for seed companies to assess new technology. In the changing seed industry scenario, seed companies in India will continue small-scale plant breeding, product testing and maintaining a strong, localized seed distribution network with cordial customer relationship. Using this strength they can get involved in alliances and agreements that may ensure access to the future technology. The future trend is quite clear from Monsanto's equity acquisition in Mahyco; Mahyco-Monsanto Biotech Ltd.; Monsanto acquiring Cargill and buying equity in E.I.D.Parry's seed business; Agr-Evo acquiring Pro-agro; Semenis buying 40 percent additional equity stock in Nath Sluis and so on. Hence the Indian seed companies will continue to play a basic and important role. Like present day contract grower farmers, small seed companies of all sizes will increasing become contract producers and developers for the bigh technology rich companies. Because of rapidly changing new technology the seed industry world over is going through a radical change with a whole new future of rapid growth ahead of it. Indian seed sector would have no option but to follow that path with active government support and huge private investments wherever needed.

A large cadre of highly skilled seed multipliers exist. Companies both in public and private sector have developed systems for training technical man power to monitor expanded infrastructure. As their R&D gains strength, they will develop new hybrids and varieties suitable not only for Indian market but also for the agro-climatic conditions prevailing in the export market. India's low cost skills at all levels of seed industry from research through seed production coupled with range of environments suitable for production of most crop seeds and rapid flow of technology offers excellent base for competitive supplies in the international market. Indian seed industry has reached a stage of maturity and is in a position to shape itself to compete at national and international levels. This may however, be recognized that private sector for some more time will continue to concentrate on more lucrative sector of the market, with a strong emphasis on hybrids and the more specialized low volume crops, e.g. vegetables. Fundamental strength and potential of Indian seed market cannot be doubted. The country has a very large population and a strong commitment to self-sufficiency in food production. This can only be achieved by increasing productivity per unit area.

High quality improved variety seed has already been accepted by the educated, well-to-do farmers. They have derived benefits by using superior genetic material and the associated technology by harvesting high yields. They realize that seed is the cheapest input in their total cost of cultivation and hence are willing to pay its fair price. They would certainly try high-tech value added seed as and when they appear in the market place and if convinced of their benefits accept them. The challenge in the coming years would be to convince the small and marginal farmers to accept quality seed of recommended, adaptable, improved hybrids of varieties. Regional private sector seed companies would probably specialize in specific farming systems to cater to the needs of such arming families in their entirety by developing area specific technology considering the economic condition and the agro-climate of the region. These companies would play a crucial role as catalyst in transferring such technologies from research labs to the farmers fields.

Indian economy is opening up the opportunities and challenges offered by the globalization of agriculture, particularly in the seed sector. Given that the prevailing uncertainites about legislative changes which currently retard seed industry growth are favourably clarified, the present trend of industry's accelerated growth in domestic market, value addition to their products, custom production and export will continue. Indian seed industry is on the threshold of becoming a global player. Beginning next century, Indian private sector can look forward to playing a key role in the development of Indian seed industry and take its rightful place in the global market.

Bibliography

Adler-Nissen, J. : *Enzymic Hydrolysis of Food Proteins,* London: Elsevier Applied Science, 1985.

Allchin, F.R.: *The Agriculture of Early Historic South Asia*, Cambridge University Press, U. K., 1995.

Bora, K.K. : *Agros Dictionary of Plant Physiology and Biochemistry,* Agrobios, Delhi, 2001.

Boyer , J.S.: *Measuring the Water Status of Plants and Soils,* Academic Press, N.Y., 1995.

Chadha K. L. and Pareek O. P.: *Advances in Horticulture: Fruit Crops,* New Delhi, Malhotra Publishing House, 1993.

Chopra, P.N.: *Agriculture of Ladakh: Our Cultural Fabric Series.* New Delhi: Ministry of Education and Social Welfare, 1978.

Dar, Ghulam Hassan : *Soil Microbiology and Biochemistry,* New India Publishing Agency, Delhi, 2010.

Devi, C.R. Sudharmai : *Analytical Procedures in Soil Science and Agricultural Chemistry,* Agrotech, Delhi, 2004.

Featherly H. I.: *Taxonomic Terminology of the Higher Plants*, USA, Iowa State College Press, 1954.

Ghosh, A.R. : *Biopesticide and Integrated Pest Management*, APH, Delhi, 2009.

Hardy B.: *Biology and Agronomy of Forage Arachis*, Cali, International Centre for Tropical Agriculture, 1994.

Herminie Broedel Kitchen: *Soils and Crops : Diagnostic Techniques*, Satish Serial Publishing, Allahabad, 2004.

John C. Benghin: *Global Agriculture Trade and Developing Countries*, Manas, Delhi, 2005.

Kapoor, R.L. and M.L. Saini: *Plant Breeding and Crop Improvement*, CBS, Delhi, 1997.

Kaul, M.K. : *Medicinal Plants of Kashmir and Ladakh : Temperate and Cold Arid Himalaya*, Indus, Delhi, 1997.

Khanna, V K : *Objective Genetics, Biotechnology, Biochemistry and Forestry,* I.K. International Publishing House, Delhi, 2008.

Madan Lal Bagdi: *Physiology, Biochemistry and Biotechnology*, Manglam Pub, Delhi, 2007.

Mistry, B.D. : *A Handbook of Spectroscopic Data : Chemistry*, Oxford University Book Co, Delhi, 2009.

Narvekar, Raghunath : *Molecular Biochemistry : Principles and Practices,* Adhyayan Pub, Delhi, 2008.

Nobel, P. S.: *Physicochemical and Environmental Plant Physiology,* Academic Press, San Diego, 1999.

Ojha, Jai Shankar : *Aquaculture and Nutrition and Biochemistry,* Agrotech, Delhi, 2006.

Pareek O. P.: *Advances in Horticulture: Fruit Crops,* New Delhi, Malhotra Publishing House, 1993.

Patel, S V and B A Golakiya : *DNA : A Bridge Between Biochemistry and Biotechnology,* New India Publishing Agency, Delhi, 2008.

Rahman, H. and A.K. Mohanty: *Entrepreneurship in Agriculture : Scopes and Opportunities*, Agrotech Pub, Delhi, 2011.

Rai, Lajpat: *Experimental Designing and Data Analysis in Agriculture and Biology*, Agrotech, Delhi, 2010.

Sharma, R K and S P S Sangha : *Basic Techniques in Biochemistry and Molecular Biology,* I K International, Delhi, 2009.

Sharma, Vijay Paul: *Glimpses of Indian Agriculture : Macro and Micro Aspects*, Academic Foundation, Delhi, 2008.

Singh, S K : *Biotechnology, Plant Propagation and Plant Breeding*, Campus Books, Delhi, 2008.

Sudharmai Devi: *Analytical Procedures in Soil Science and Agricultural Chemistry*, Agrotech, Delhi, 2004.

Swaminathan, MS : *From Green to Evergreen Revolution : Indian Agriculture: Performance and Challenges*, Academic Foundation, Delhi, 2010.

Swarnim, K. : *A Textbook of Biochemistry and Microbiology,* Surendra Pub, Delhi, 2010.

Tyagi, I.D. : *Plant Breeding and Genetics at a Glance*, South Asian, Delhi, 2005.

Vanangamudi, K : *Principles and Methods of Plant Breeding*, International Book, Delhi, 2005.

Verma, L.R. and R.C. Sharma: *Diseases of Horticultural Crops: Fruits*, Indus, Delhi, 1999.

Vijay Paul Sharma: *Glimpses of Indian Agriculture : Macro and Micro Aspects*, Academic Foundation, Delhi, 2008.

Yadav, M : *Nutritional Biochemistry and Metabolism,* Arise Pub, Delhi, 2008.

Yoshida T.: *Cultivation of Citrus Genetic Resources for Evaluation of Characteristics*, Tokyo, JICA, 1996.

Index

H

I

K

M

N

P

R

S

T

V

W

❑❑❑